Edited by
Klaus Ammann
Yolande Jacot
Gösta Kjellsson
Vibeke Simonsen

Methods for Risk Assessment of Transgenic Plants

III. Ecological risks and prospects of transgenic plants, where do we go from here? A dialogue between biotech industry and science

Springer Basel AG

Editors:

Dr. Klaus Ammann
Dr. Yolande Jacot
Botanical Garden
Altenbergrain 21
CH-3013 Bern
Switzerland

Dr. Vibeke Simonsen
Dr. Gösta Kjellsson
Department of Terrestrial Ecology
National Environmental Research Institute
Vejlsøvej 25
DK-8600 Silkeborg
Denmark

A CIP catalogue record for this book is available from the Library of Congress, Washington D.C., USA

Deutsche Bibliothek Cataloging-in-Publication Data

Methods for risk assessment of transgenic plants. - Basel ; Boston :
Berlin : Birkhäuser

3. Ecological risks and prospects of transgenic plants, where do we
go from here? : a dialogue between biotech industry and science /
ed. by Klaus Ammann ... - 1999
Bd. 1. verf. von Gösta Kjellsson und Vibeke Simonsen. -
ISBN 978-3-0348-9735-8 ISBN 978-3-0348-8700-7 (eBook)
DOI 10.1007/978-3-0348-8700-7

Originally published by Birkhäuser Verlag in 1999
Softcover reprint of the hardcover 1st edition 1999
Printed on acid-free paper produced from chlorine-free pulp. TCF ∞

ISBN 978-3-0348-9735-8

9 8 7 6 5 4 3 2

Table of contents

Session 3: Short-term, long-term effects and standardisation of limits

Session 4: Monitoring methods

Session 5: Population genetics

Session 6: Decision procedures, harmonisation

Session 7: Methodological lacunas

Session 8: Conclusion, strategies, where do we go from here?

Methods for Risk Assessment of Transgenic Plants
III. Ecological risks and propspects of transgenic plants
K. Ammann, Y. Jacot, V Simonsen and G. Kjellsson (eds)

Setting the scene I

Gösta Kjellsson

National Environmental Research Institute, Denmark

Dear colleagues, ladies and gentlemen.

Two reasons for having this conference are the two method books published by Birkhäuser Verlag. In 1991, I was asked together with Vibeke Simonsen by the Danish Environmental Authorities to prepare a catalogue of methods for risk assessment of transgenic plants. This work should cover essential aspects in relation to competition, plant establishment and ecosystem effects.

An extensive literature search was made, and during some intense months of work, the first version was produced. After the result was presented to our colleagues in the Ministry, it was decided that it should be published as a book. We contacted Birkhäuser Verlag in Basel, and the first book was published in 1994. (Kjellsson and Simonsen, 1994)

Already during the preparation, it became clear that a second volume was greatly needed, with focus on: Pollination, gene transfer and effects on plant populations. It was also clear that neither Vibeke or myself had the specific professional skills to cover this vast area. In the meantime we had come into contact with Klaus Ammann here in Bern, and the Swiss group which were engaged in this area. The Swiss group was interested in collaboration with Danish scientists – and consequently, a plan was made and a project was set up.

The project was supported by the Federal Office of Environment, Forest and Landscape, Switzerland, the European Commission, DG XI, and from Denmark: the National Forest and Nature Agency, and the Environmental Protection Agency.

The project involved many discussions concerning the scientific basis, the use of terminology and how to define and apply different types of methods to specific topics in risk assessment.

Many of these questions were solved during two fruitful workshops, one in Switzerland and one in Denmark.

During the last stage of the work, it was decided that the publishing of the method books should be followed by an international conference on important aspects of risk assessment, to get an even broader view. (Kjellsson et al., 1997)

This was the seed of the conference we are starting here today. For different reasons – some of which are local – the conference is being held here in the beautiful town of Bern, Switzerland – and I suspect that not many of you mind that at all.

We have a large program ahead of us, and many exciting topics for discussion, so I will and now give the platform to my dear colleague, Klaus Ammann.

Kjellsson Gösta, Vibeke Simonsen (1994) Methods for Risk Assessment of Transgenic Plants I. Competition, Establishment and Ecosystem Effects. Birkhäuser, Basel, 214

Kjellsson Gösta Vibeke Simonsen, Klaus Ammann (1997) Methods for Risk Assessment of Transgenic Plants II. Pollination, Gene transfer and Population impacts, Birkhäuser, Basel, 308

Methods for Risk Assessment of Transgenic Plants
III. Ecological risks and propspects of transgenic plants
K. Ammann, Y. Jacot, V. Simonsen and G. Kjellsson (eds)

Setting the scene II

Klaus Ammann

Botanical Garden of the University of Bern, Switzerland

Welcome to Bern, my friends, and welcome to the "Gene Monastery" in our Botanical Garden. This I say on purpose, since this will not be an ordinary symposium where everybody tries to show how perfect and intelligent his own contribution to Science is, this will be a discussion process, a collaborative learning process. It is set up as a "trialogue" between Scientists, Regulators and Industrialists in a new concept: Scientists will set the tone, they will have a voice to contribute to each of the modules described later – and Regulators and Industrialists have to listen. At the end of each module Regulators and Industrials are obliged to speak up, to intervene, and to do this with an open mind and with no fear about revealing governmental and corporate interests. The symposium is set up in a way that corporate people should feel free to participate, in an atmosphere of collaborative thinking, and for once not feel censured by environmentalists.

For this purpose, modules have been set up in a specific sequence, in order to encourage a learning process, which will end up in some resolutions, resolutions which can be supported by everyone attending symposium. This is also the reason why we have avoided to open the symposium to a broader audience, since we want to initiate a dialogue between professionals on risk assessment. Let me come back to the opening remark about the "Gene Monastery". Catering, pauses and even the evening meals will be organised within the Botanical Garden, so we have a unique chance to concentrate with complete focus.

We will start with *Ecological Effects of Transgenes*, a module which will set the tone, since the title of the symposium involves questions about ecological effects and prospects. The study of ecological effects can be considerably helped by *Modelling Risk Assessment*, and inevitably we will dig into the problematic of *Short-term and Long-term Effects* and also into the important questions about a possible *Standardisation of Limits.* The first symposium day will finish with a more synthetic module about *Monitoring*, which will allow us to wrap up in a first phase what has been said and discussed.

With fresh minds the next day, we will tackle the complex science of population genetics, another basic view whereby we analyse ecological risk assessment of transgenic crops.

Then the symposium inevitably will sail into the political waters of *Decision Procedures and Harmonisation.* Having done this we will have a good basis to attack the very important questions about the *Methodological Lacunas. Conclusions, Strategies* will then be the final topic of a module, which will hopefully reveal some information on *Where Do We Go from Here?*

As you can easily see, this is a very demanding task to be pragmatic, we will not be able to solve all the problems that will occur in these times of a first mass dissemination of transgenic crops on a global scale. But as one of the organizers, I have strong hopes that this will be a good beginning for future development in risk assessment, which takes more into account long-term ecological problems.

Session 1: Ecological effects of transgenes

Methods for Risk Assessment of Transgenic Plants
III. Ecological risks and propspects of transgenic plants
K. Ammann, Y. Jacot, V. Simonsen and G. Kjellsson (eds)

Predicting the ecological impacts of transgenes for insect and virus resistance in natural and feral populations of *Brassica* species

Alan F. Raybould[1], Catherine L. Moyes[2], Lindsay C. Maskell[1], Rebecca J. Mogg[1], Elizabeth A. Warman[3], Judith C. Wardlaw[1], Graham W. Elmes[1], Mary-Lou Edwards[4], J. Ian Cooper[4], Ralph T. Clarke[1] and Alan J. Gray[1]

[1] *Natural Environment Research Council, Institute of Terrestrial Ecology, Furzebrook Research Station, Wareham, Dorset BH20 5AS, UK*
[2] *Present address: Brassica and Oilseeds Research Department, John Innes Centre, Colney Lane, Norwich NR4 7UH, UK*
[3] *Natural Environment Research Council, Institute of Terrestrial Ecology, Monks Wood Abbots Ripton, Huntingdon, Cambridgeshire PE17 2LS, UK*
[4] *Natural Environment Research Council, Institute of Virology and Environmental Microbiology, Mansfield Road, Oxford OX1 3SR, UK*

Introduction

Oilseed rape is now widely grown in the UK. Nearly 400 000 ha were planted in 1990, compared with only 50 000 ha in 1978 [1]. There is concern that transgenes may be able to escape from cultivated oilseed rape, either by the formation of feral populations that arise from spilled seed [2, 3], or by gene flow into related wild species (including *B. oleracea*) [4, 5]. This paper describes research to help predict whether transgenes for insect or virus resistance (two of the commonest genetically modified stress-tolerance traits [6]) will increase the weediness of feral and wild *Brassica* populations in the UK. We also consider resistance to herbivory by molluscs (although we know of no genetic modifications for mollusc resistance) because slugs and snails are common herbivores of *Brassica* species.

Predicting the effects of transgenic insect and virus resistance in wild *Brassica oleracea*

Variation for glucosinolates in Dorset populations of B. oleracea

Plant populations are often polymorphic for insect resistance. It is assumed that variation in resistance occurs, at least in part, because genetically controlled resistance mechanisms impose a cost in the absence of attack (e.g. [7]). It is possible that GM insect resistance may impose no cost [8] and, therefore, could spread rapidly within and among populations and possibly alter their dynamics.

To estimate the potential effects of herbivore resistance transgenes on the population dynamics of a wild crop relative, we have studied the distribution of genetic variation controlling the production of glucosinolates in natural populations of *Brassica oleracea* (wild cabbage) on the Dorset coast. Glucosinolates are secondary metabolites that occur in the Capparales and a few unrelated taxa. They are composed of a glycone group with a variable side-chain derived from methionine (aliphatic glucosinolates), phenylalanine (aromatic glucosinolates) or tryptophan (indolyl glucosinolates). Aliphatic glucosinolates are the most abundant type in *Brassica* leaves. When tissues are damaged, the endogenous enzyme myrosinase is released which, under ambient temperature and pH, calatyses the hydrolysis of aliphatic glucosinolates to isothiocyanates (mustard oils) (e.g.[9]).

In oilseed rape, variation in glucosinolate concentration and side-chain structure mediate interactions with herbivores. For example, high concentrations deter feeding by generalist herbivores such as rabbits, pigeons and slugs while attracting and stimulating feeding and egg-laying in specialist *Brassica* herbivores such as flea-beetles. The effects on specialists are increased by shortened side-chain structure (propenyl rather than butenyl glucosinolates) and decreased hydroxylation [10, 11].

Interactions between herbivores and aliphatic glucosinolates in wild cabbage are an attractive system for studying the potential impacts of insect resistance transgenes on natural populations. Variation in side-chain modification of aliphatic glucosinolates in *B. oleracea* is under the control of a small number of loci [12] and aliphatic glucosinolate concentration appears to have high heritability [9]. The seedling and pollination biology of wild cabbage and feral oilseed rape are also very similar, although cabbage is a long-lived perennial whereas oilseed rape is annual or biennial [13, 14].

Mithen et al. [9] compared variation at loci controlling glucosinolate side-chain structure with that at isozyme loci in cabbage plants at three sites on the Dorset coast. There was strong differentiation (high and significant F_{ST} values) at all glucosinolate loci but not at isozyme loci (low and non-significant F_{ST}). There were also clear differences in aliphatic glucosinolate concentrations in leaf tissue, which was also reflected in glasshouse-grown seed material taken from the same sites. Mithen et al. inferred from these data that the glucosinolate differences were not the result of genetic drift or founder effects, and suggested divergent selection among sites was a possible cause.

C.L. Moyes and A.F. Raybould (unpublished observations) have studied genetic variation at glucosinolate and marker loci in more detail. Leaf tissue was sampled from 50 plants, spaced 1 m apart in each of seven transects. At Durdle Door, Kimmeridge and St. Aldhelm's Head (Fig. 1) two transects were sampled to reflect environmental variation (mainly substrate and vegetation cover) within the site. At Old Harry a single transect was sampled. The leaf material was assayed for aliphatic glucosinolate phenotype and concentration using HPLC. The frequency of null homozygotes *pro*, *elong* and *oh* loci was inferred from the chromatograms and allele frequencies were estimated from genotype frequencies using a maximum likelihood procedure [15]. The proportion of the variance in allele frequencies due to site effects was estimated using analysis of molecular variance (AMOVA) [16], with groups of 12 plants as the unit of comparison. Glucosinolate concentration was analysed by hierarchical ANOVA using the same groups. Two-hundred-and-fifty additional plants were sampled from the same sites for analysis with six microsatellite primer pairs isolated from oilseed rape [17, 18]. Data were also analysed by AMOVA using both differences in allele states (F_{ST}) and the sum of the squared allele size differences (R_{ST}) [19] with spatially distinct groups of about 10 plants as the unit of comparison. The plants were also analysed for variation in esterase, malate dehydrogenase and peroxidase isozymes (eight putative loci).

There was a significant among population effect for all loci or groups of loci analysed (Tab. 1). Compared with the results of Mithen et al. [9], the data from a wider range of populations show a much less clear difference between the distribution of variation at marker and glucosinolate loci. There is strong differentiation among populations at the *oh* locus, but *pro*

Figure 1. *B. oleracea* study populations.

Table 1. Partitioning of genetic variation at microsatellite, isozyme and glucosinolate loci and for TuMV resistance within and among 5 Dorset populations of *B. oleracea*

Marker/ trait	Percentage of variation		
	Within patches	Among patches within populations	Among populations
Microsatellites (R_{ST})	84.15	−0.55	16.40
Microsatellites (F_{ST})	75.92	1.27	22.81
Isozymes	77.38	14.14	8.48
Pro	79.74	15.60	4.66
Elong	70.78	18.13	11.09
Oh	52.89	2.62	44.49
Glucosinolate concentration	83.36	7.72	8.92
TuMV resistance*	83.70	12.67	3.63

*Within half-sib families/among families within populations/among populations.

and *elong* show a similar amount of differentiation to isozymes and lower differentiation than microsatellites. Therefore while there are statistically significant differences in allele frequencies at glucosinolate loci and in glucosinolate concentrations among populations, it is not possible on the basis of these data to infer that genetic drift and limited gene flow cannot fully account for the differences.

Analysis of among population variation, however, does not examine whether the spatial pattern of variation is different at glucosinolate and marker loci. Preliminary analysis (Moyes, unpublished observations) suggests that glucosinolate and marker variation may be distributed differently within populations. Glucosinolates tend to be uniform within transects but different between transects within populations, whereas markers tend to show as much variation between groups within transects as between groups between transects within sites. The situation is complex, however, as microsatellites show very little differentiation within populations (among patches within populations column in Tab. 1). All other data suggest that populations are genetically substructured; isozymes show significant F_{IS} (Tab. 1 and [9]), there is non-random distribution of cleaved amplified polymorphic DNA variation at self-incompatibility locus-related sequences [20] and direct observations of pollinator flights and seed dispersal suggest that genetic neighbourhoods contain fewer than 20 individuals (Warman, unpublished observations). The lack of detectable within population differentiation may be due to higher mutation rates at microsatellite loci compared with other markers (e.g. [21]).

Associations between glucosinolates and herbivore damage

Although the genetic structure work is equivocal about whether there is selection at glucosinolate loci, glucosinolate variation does appear to be associated with variation in potentially important life-history characters. Moyes et al. [13] grew mixtures of seedlings from St. Aldhelm's Head (large proportion of plants with high concentrations of 3-butenyl glucosinolate) and Kimmeridge (most plants with low concentrations of a mixture of 2-propenyl, 3-butenyl and 2-hydroxy 3-butenyl glucosinolates) at each of the two sites in the summer of 1995. At both sites, a higher proportion of seedlings originating from St. Aldhelm's Head were grazed by molluscs (Kimmeridge $\chi^2 = 19.98$, d.f. = 1, $p < 0.001$; St. Aldhelm's Head $\chi^2 = 32.19$, d.f. = 1, $p < 0.001$). Flea beetles occurred at Kimmeridge only, where they attacked a higher proportion of seedlings originating from St. Aldhelm's Head ($\chi^2 = 14.30$, d.f. = 1, $p < 0.001$). Therefore the generalist *Brassica* herbivores (molluscs) attacked a higher proportion of the seedlings from the population with low concentrations of 3-butenyl glucosinolate, whereas seedlings from the population with high concentrations of 3-butenyl glucosinolate were attacked more frequently by the specialist flea beetles. This is what one would expect on the basis of data from experimental populations of oilseed rape [11]. However, near isogenic lines differing only in glucosinolate profiles, would be required to confirm that the glucosinolate differences are the basis of the herbivore preferences.

There also appears to be an association between glucosinolate variation and fitness of mature plants. While there is no detectable effect of herbivory of mature leaves on seed output [13], there is an effect of herbivory of seed pods. Cabbage seed pods are attacked by larvae of the cabbage seed weevil (*Ceutorhyncus assimilis*) and several studies have shown that weevils are attracted by isothiocyanates [22–25]. Seed loss by direct consumption is exacerbated by the entry of fungal pathogens through the damaged pod wall [26]. Damage to the pod wall also facilitates egg-laying by the *Brassica* pod midge (*Dasineura brassicae*) which can lay eggs in intact pods only with difficulty.

Moyes and Raybould [26] found significant differences among transects in the proportion of pods infested with cabbage seed weevil in both 1995 and 1996 ($\chi^2 = 65.58$, d.f. = 6, $p < 0.001$ and $\chi^2 = 174.23$, d.f. = 6, $p < 0.001$ respectively). In both years, transects at St. Aldhelm's Head (i.e. high 3-butenyl glucosinolate) had the highest proportions of infested pods, while plants at Kimmeridge had the lowest proportions. In addition, these transects also had the highest infestations of brassica pod midge (Fig. 2).

The effect of weevil infestation on seed output was examined at St. Aldhelm's Head in 1996 [26]. There was no relationship between the proportion of pods infested and the number of pods per plant (r = 0.10, $p = 0.379$), the proportion of pods infested and plant size measured as leaf area (r = 0.14, $p = 0.23$) or the number of racemes per plant (r = 0.12, $p = 0.31$). A weighted regression of the mean number of seeds per pod against the proportion of pods infested showed that seed output was significantly reduced by higher amounts of weevil infestation (r = 0.60, $p = 0.05$) [26].

The preliminary work presented here suggests that glucosinolate variation has detectable effects on damage caused by herbivores. Seedlings with high concentrations of 3-butenyl glucosinolate seem more prone to flea beetle damage and less prone to mollusc damage than seedlings with lower concentrations. Also populations with high proportions of 3-butenyl glu-

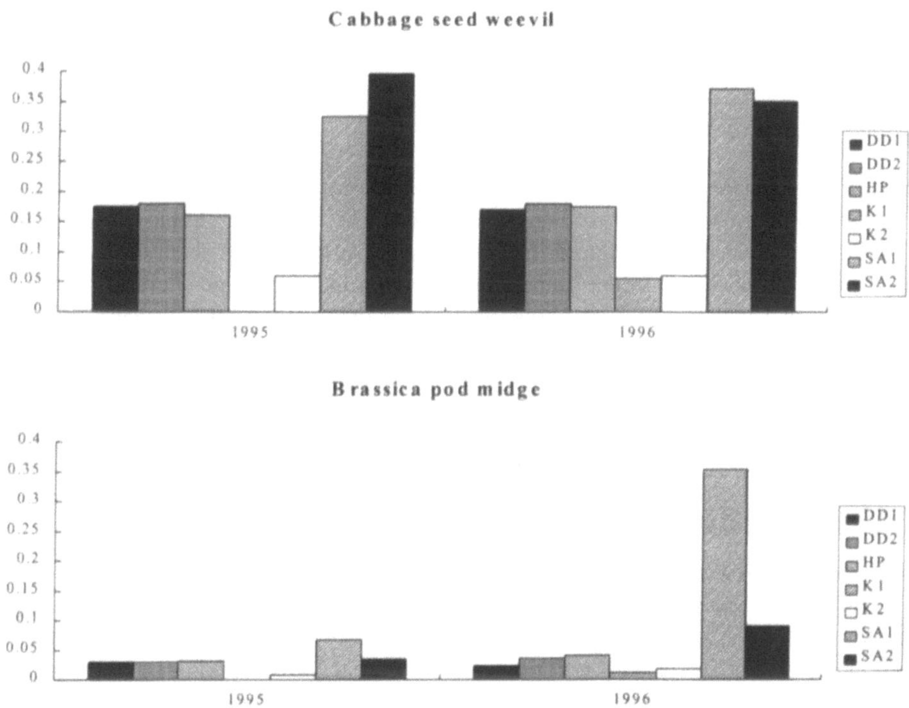

Figure 2. The proportion of pods infested with cabbage seed weevil and brassica pod midge in 7 transects of *B. oleracea*.

cosinolate suffer higher infestations of cabbage seed weevil and brassica pod midge, and weevil infestation appears to be associated with reduced seed output [26] (although there is no association between weevil infestation and concentration of 3-butenyl glucosinolate at the individual plant level – C.L. Moyes, unpublished observations).

We cannot, however, infer from these data that the glucosinolate polymorphism originated or is maintained by fitness trade-offs due to the effects of specialist and generalist herbivores. Nor, by extension, can we argue that a general insect (or herbivore) resistance transgene would significantly alter the fitness of wild cabbage plants. Although slugs appear to selectively browse seedlings with lower glucosinolate concentrations, and although cabbage populations with high glucosinolate concentrations have high weevil infestations, seedling recruitment is rare in wild cabbage populations [27] and as yet we have no evidence that recruitment is other than random with respect to glucosinolate profiles. Therefore, while glucosinolates may influence herbivore feeding behaviour, their effects on plant fitness may be minimal.

The other important point to draw from the work on herbivory of wild cabbage is the complexity of interactions between herbivores, parasitoids and diseases. Weevil damage in terms of lost seeds per pod is reduced by parasitoid wasps (as has also been demonstrated in the interaction between *Hormathophylla spinosa* (Cruciferae) and a *Ceutorhyncus* species [28]), whereas pod midge infestation and fungal disease increase with weevil infestation [26] and all these factors may interact with glucosinolate phenotype. If the interaction between glucosinolates and herbivores is in any way analogous to that between herbivores and insect resistant transgenes, the effect of such transgenes in natural populations will (perhaps obviously) depend on the outcome of the complex dynamics of multispecies interactions, rather than the relationship between the plant and a single herbivore.

Ecological effects of virus resistance transgenes

One concern about the release of plants containing virus resistance transgenes is that the genes will introgress into a wild relative of a crop and cause it to become weedy as the effects of viruses on fitness are removed or reduced [6, 29]. There are also concerns about the evolution of new virus types which may result from recombination between pathogen-derived virus resistance genes and viruses [30, 31]. Assessment of the risks associated with these processes requires knowledge of the distribution of viruses in populations of crop relatives and their effects on the population dynamics of the species. To investigate these issues, we are studying the growth and reproduction of wild *Brassica oleracea* in relation to virus infection.

Distribution of viruses in natural populations of Brassica oleracea

Two-hundred-and-eleven plants among five populations were assayed for the presence of four virus pathotypes (beet western yellows virus, BWYV; cauliflower mosaic virus, CaMV; turnip mosaic virus, TuMV; and turnip yellow mosaic virus, TYMV). ELISA [32] revealed significant non-random distribution of all four virus types (Tab. 2), the most striking variation being in TYMV, in which 82% of the plants with the virus occurred in a single population. Skotnicki et al. found a similar extremely non-random distribution of TYMV in *Cardimine lilacina* on Mt. Kosciusko in Australia [33].

Table 2. The distribution of viruses among 5 Dorset populations of *B. oleracea*

	Plants tested	Plants with BWYV	Plants with CaMV	Plants with TuMV	Plants with TYMV
Old Harry	50	16%	26%	36%	2%
Winspit	44	55%	80%	57%	0%
Chapman's Pool	50	52%	24%	90%	62%
Kimmeridge	30	67%	50%	80%	13%
Durdle Door	37	35%	41%	41%	5%
Test for random distribution					
χ^2		15.57	20.93	16.66	72.43
d.f.		4	4	4	4
P		0.0044	0.0003	0.0023	<0.0001

Associations among viruses were also non-random. There were more plants than expected with zero or three virus types and fewer than expected with one or two virus types ($\chi^2 = 62.32$; d.f. = 4; $p < 0.00005$). Tests of associations among pairs of viruses (Tab. 3) showed that all viruses occurred together with CaMV more often than expected. There was also a significant positive association between BWYV and TuMV. The associations between BYMV, CaMV and TuMV may be because they share the peach potato aphid (*Myzus persicae*) as a vector and CaMV and TuMV are also spread by the cabbage aphid (*Brevicorynae brassicae*) [34]. TYMV is spread by biting insects, such as weevils and flea beetles [34], so it is not surprising that there is no association between the distributions of BWYV and TYMV. A possible reason for the non-random distributions of TYMV with CaMV and TuMV may be common susceptibility to TYMV and CaMV and cross-protection or fitness trade-offs in the case of TYMV and TuMV.

The reason for the observed non-random distribution of viruses may be genetic differences among plants (populations) or the distribution and behaviour of virus vectors. The available evidence suggests vector behaviour as the more important. A study of resistance to turnip mosaic virus (Cooper, Edwards and Raybould, unpublished observations) found a significant heritable component to variation in TuMV titres among plants in a glasshouse trial (h^2 among Chapman's Pool, Winspit and Kimmeridge populations = 15.5–31.0%, variation among families $F = 2.82$; d.f. 57, 660; $p < 0.0001$), but only a small proportion of the variation was due to differences among populations (Tab. 1) (although among population variation was significant $F = 4.64$; d.f.

Table 3. Tests for association between virus types

	BWYV	TuMV	CaMV
TuMV	$\chi^2_{[1]} = 15.894$ $p = 0.000$ +ve		
CaMV	$\chi^2_{[1]} = 36.366$ $p = 0.000$ +ve	$\chi^2_{[1]} = 9.483$ $p = 0.002$ +ve	
TYMV	$\chi^2_{[1]} = 1.707$ $p = 0.191$ no association	$\chi^2_{[1]} = 8.841$ $p = 0.003$ –ve	$\chi^2_{[1]} = 13.740$ $p = 0.000$ +ve

2, 57; $p = 0.0136$). In addition, control plants in a field trial (see below) were infected with viruses in roughly equal proportions regardless of their origin.

The effects of viruses on plant fitness

To test for the effects of viruses on plant fitness, 574 seedlings from the five study populations were grown in insect-screened glasshouses in Oxford. At the three to five leaf-stage, plants were inoculated with either TuMV (an isolate from Kimmeridge), TYMV (an isolate from Chapman's Pool) or sterile water. When the plants were 3 months old (June 1996) they were planted out in a fully-randomised design in a field at Chapman's Pool, approximately 300 m away from a natural population of *Brassica oleracea*.

Mortality was significantly non-random among the treatments ($\chi^2_{[2]} = 17.29$, $p < 0.001$), with highest mortality in the TYMV treatment and lowest in the TuMV (Fig. 3). Among the surviving plants there were also differences in (assumed) fitness characters (Maskell, unpublished observations). For example there were significant differences in total seed production per plant among the treatments ($F = 5.00$, d.f. = 2. 79; $p = 0.009$), with the virus infected plants producing fewer seeds than the control plants (Fig. 4).

Our results show that agriculturally important viruses are present in natural populations of *B. oleracea* and that infection with these viruses can affect plant fitness, especially if infection

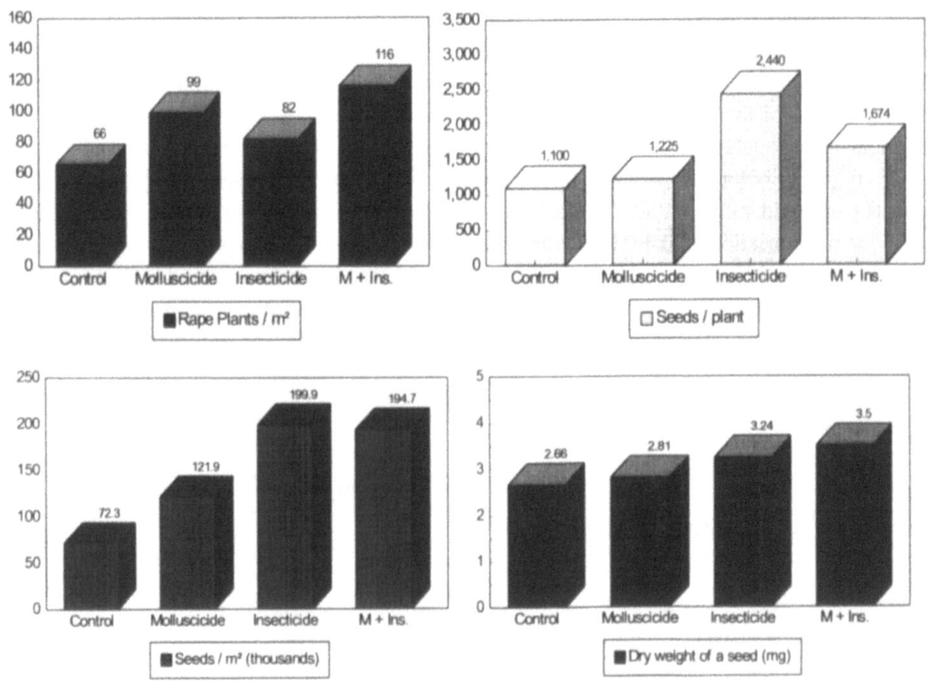

Figure 3. Percent mortality of cabbage plants in the field trial at Chapman's Pool.

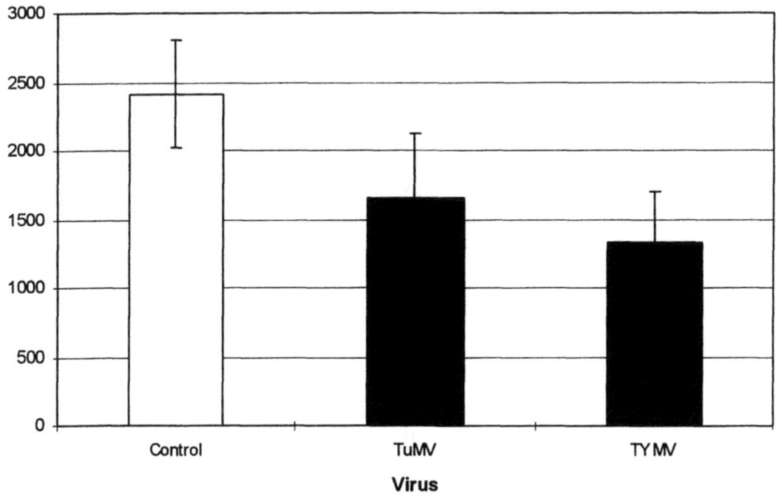

Figure 4. Seed production per plant (± SE) in the field trial at Chapman's Pool.

occurs at the seedling stage. This contrasts with the results of Bartsch et al. [35] who found that sugar beet resistant to beet necrotic yellow vein virus (BNYVV) produced more biomass than susceptible beet in trials on a site infected with BNYVV. Transgenic BNYVV-resistant beet gave intermediate productivity. However, no BNYVV was detected in wild beets in northeast Italy, suggesting that introgression of gene for BNYVV resistance would have no effect on wild beets in that area.

Genetic differences among the *B. oleracea* populations appear to play only a small part in the observed non-random distribution of viruses. This suggests that vector behaviour may be the major determinant of virus distribution, and that vector distributions are not consistent enough to impose differential selection for virus resistance among populations. This contrasts with the apparent consistency of cabbage seed weevil distributions (see above). However, wild cabbage has a long life-cycle and consistent vector patterns over a few years may be insufficient to impose differential or directional selection. Clearly, therefore, the effect of a virus resistance transgene in a natural population will be highly dependent on the behaviour of the vectors of the virus against which the gene provides resistance.

Predicting the effects of transgenic insect and mollusc resistance in feral populations of oilseed rape

Little is known of the importance of herbivory in regulating the size and persistence of feral populations. To simulate the effect of transgenes for resistance to insect and/or mollusc herbivory we decided to apply insecticide and/or molluscicide to feral populations. However, surveys of feral rape populations (Wardlaw, unpublished observations) indicated that it was not possible to conduct well-replicated experiments because of the transience of populations (e.g. because of mowing), varietal differences, varying distances from herbivore source populations and variation in population size. We therefore established experimental populations of "feral" rape on farmland near Moreton, Dorset.

Experimental plots were established on pasture that had been ploughed and harrowed before the start of the experiment. Briefly, the experiment was set out as follows. Twenty-five 5 m × 5 m plots were marked out in five rows of five. On 28th March, seed of spring rape variety Aries was sown into the central 2 m × 2 m part in 17 rows at a density calculated to give 35 plants per row (previous trials indicated 86% germination) (i.e. about 150 plants m^{-2}). Molluscicide ("Draza"), insecticide (alternately "Hallmark" and "Sybol") or both were applied periodically to individual plots according to the manufacturers' recommendations according to a latin square design. The 2 × 2 m plots were fenced to prevent grazing by deer and rabbits and the 1.5 m surrounding strips were kept free of weeds until seeded naturally by the mature rape plants. Among several performance measures, the density of plants in each plot was recorded on 25th July and seed output was recorded on 14th August.

Molluscicide and insecticide treatments both increased the density of plants over that of the control plots (Fig. 5), although only the molluscicide effect was significant at the 5% level. Combined molluscicide and insecticide treatment gave the highest plant density. Higher plant densities with insecticide were due to reduced damage by flea beetles (e.g. *Phyllotreta* spp.).

The number of seeds per plant was increased significantly by insecticide but not by molluscicide. The combined treatment gave a value intermediate between the two single treatments (Fig. 5). Several effects combined to give these results. The small difference between the control and molluscicide plots was probably due to lower plant density in the controls, leading to the production of larger plants. This compensated for damage caused by molluscs. The insecticide treatment had a large effect because it reduced damage to flowers by pollen beetles (*Meligethes* spp.) and seed predation by the cabbage seed weevil (*Ceutorhyncus assimilis*). Plants also flowered earlier than in the other plots because they were not checked by flea bee-

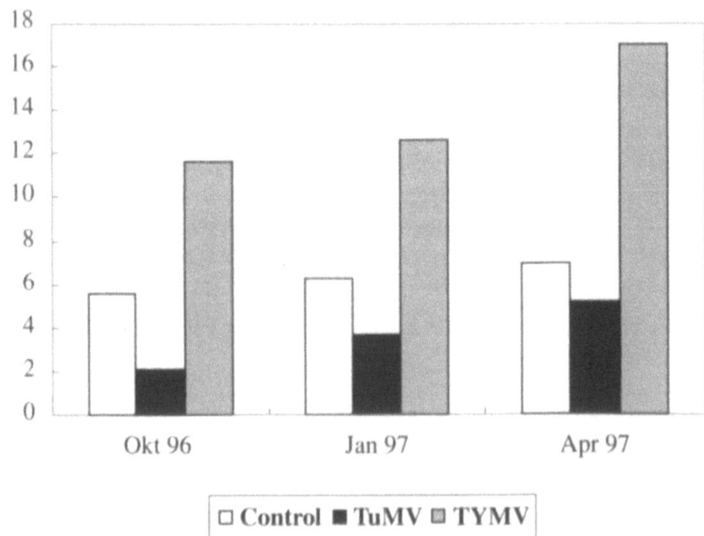

Figure 5. Measures of plant performance in the oilseed rape experiments at Hurst Farm.

tle damage. The intermediate value for the combined treatment is probably because the high plant density gives smaller plants compared with the insecticide treatment alone.

The plant density and seeds per plant values combine to give an estimate of seed output per m^2. They were significant differences among the treatments ($F = 6.88$; d.f. = 4, 12; $p = 0.004$) which were due mainly to the effect of insecticide ($F = 19.52$; d.f. = 1, 21; $p < 0.0001$). Neither the molluscicide effect nor the insecticide/molluscicide interaction were significant ($F = 0.77$ and 1.24 respectively; d.f. = 1, 21; $p > 0.05$ in both cases).

These data suggest that transgenic insect resistant feral oilseed rape may produce more seed per unit area than the equivalent unmodified variety. However, this does not mean that insect resistant feral oilseed rape would necessarily cover a larger area or be more persistent than non-resistant varieties. Rape plants were allowed to seed into the 1.5 m cleared border area around each plot which received the same treatment as the 2 × 2 m central square. Significantly more seedlings were produced in the treatment plots compared with the control plots, although absolute numbers were small. On average, the control plots produced six plants from the estimated output of 250 000 seeds per plot; insecticide treated plots produced about 50 plants from 750 000 seeds; molluscicide treated plots gave about 55 plants from 500 000 seeds; and the combined treatment produced about 90 plants from 750 000 seeds.

All seedlings in the protected central 2 × 2 m plot were killed by frost, as were the smaller seedlings in the unprotected outer area. The more mature plants in the outer area were heavily predated by vertebrate herbivores (pigeons, pheasants, rabbits and especially deer). Only 36 flowering rape plants were found throughout the whole experimental site in the summer of 1997, the majority of which were in one control plot that, by chance, had been missed by the deer.

Conclusions

Observations of natural populations and field trial data suggest that herbivory and virus infections can affect plant survival and/or seed output. It is not clear, however, whether herbivory or viruses play any role in determining the persistence or abundance of natural or feral *Brassica* populations. *B. oleracea* has very high seed output compared with seedling recruitment and also has short seed dispersal distances. Many seedlings appear to die because of drought [27], and while slugs prefer seedlings with low glucosinolate concentrations if given a choice, they will eat those with high glucosinolate concentrations [13]. Similarly in feral rape populations, competition with perennial plants and herbivory by vertebrates seems far more important to population persistence than does protection from insect and mollusc herbivory. These preliminary results are similar to those of Bergelson [36], who found that a herbicide resistant line *Arabidopsis thaliana* produced fewer seed than a susceptible line in the absence of herbicide. However, there was no difference in the invasiveness of the two lines because space rather than seed production limited recruitment. Similar process may operate in *Brassica* populations, either giving no selective advantage to insect and virus resistance, or applying soft rather than hard selection. In the latter scenario, transgenes for resistance may spread among populations, but not increase plant weediness. More research is required to assess the crucial factors affecting the seedling recruitment of *Brassica* species before we can predict with confidence the effects of transgenes on their population dynamics.

The other important consideration is whether natural pest and disease resistances are useful analogues of transgenic resistance. As mentioned above, there may be trade-offs between resistance and fitness in the absence of pests or pathogens in natural populations [7], while there may be no cost to genetically modified resistance [8]. Also pests and their parasitoids may have evolved to use natural resistance mechanisms, such as glucosinolates, as cues for locating their hosts, whereas such interactions with transgenic resistance mechanisms are less likely. We also need to be careful that phenotype similarities are considered at both the whole plant and molecular levels, and under a range of conditions. For example, the robustness of transgenic and natural virus resistance mechanisms may vary with temperature in different ways [37, 38]. Transgenic virus resistance mechanisms will also interact with the pathogen in very different ways, with potentially significant effects on the evolution of new viral strains which cannot be predicted easily by studies of natural populations.

References

1. Raybould AF, Gray AJ (1994) *Genetically modified organisms research report no.1. genetically modified crops and their wild relatives – a UK perspective*. Department of the Environment, London. 124 pp + annexes
2. Crawley MJ, Brown SL (1995) Seed limitation and the dynamics of feral oilseed rape on the M25 motorway. *Proc R Soc Lond B* 259: 49–54
3. Charters YM, Robertson A, Wilkinson MJ, Ramsay G (1996) PCR analysis of oilseed rape cultivars (*Brassica napus* ssp. *oleifera*) using 5'-anchored simple sequence repeat primers. *Theor Appl Genet* 92: 442–447
4. Scheffler JA, Dale PJ (1994) Opportunities for gene transfer between from transgenic oilseed rape (*Brassica napus*) to related species. *Transgenic Res* 2: 356–364
5. Mikkelsen TR, Andersen B, Jørgensen RB (1996) The risk of crop transgene spread. *Nature* 380: 31
6. Cooper JI, Raybould AF (1997) Transgenes for stress tolerance: consequences for weed evolution. *Proceedings of the 1997 Brighton Crop Protection Conference – Weeds*, 265–272
7. Fineblum WL, Rausher MD (1995) Tradeoff between resistance and tolerance to herbivore damage in a morning glory. *Nature* 377: 517–520
8. Hilder VA, Gatehouse AMR (1991) The phenotypic costs to plants of an extra gene. *Transgenic Res* 1: 54–60
9. Mithen R, Raybould AF, Giamoustaris A (1995) Divergent selection for secondary metabolites between wild populations of *Brassica oleracea* and its implications for plant-herbivore interactions. *Heredity* 75: 472–484
10. Mithen R (1992) Leaf glucosinolate profiles and their relationship to pest and disease resistance in oilseed rape. *Euphytica* 63: 71–83
11. Giamoustaris A, Mithen R (1995) The effect of modifying the glucosinolate content of the leaves of oilseed rape (*Brassica napus* ssp. *oleifera*) on its interaction with specialist and generalist pests. *Ann Appl Biol* 126: 347–363
12. Giamoustaris A, Mithen R (1996) Genetics of aliphatic glucosinolates. IV. Side-chain modification in *Brassica oleracea*. *Theor Appl Genet* 93: 1006–1010
13. Moyes CL, Collin HA, Raybould AF (1999) The role of glucosinolates in plant-herbivore interactions in wild cabbage. *In*: AJ Gray, F Amijee, CJ Gliddon (eds): *Environmental Impact of Genetically Modified Crops*. Department of the Environment, Transport and the Regions, London, 175–187
14. Warman EA (1998) Gene flow in wild cabbage. *In*: AJ Gray, F Amijee, CJ Gliddon (eds): *Environmental Impact of Genetically Modified Crops*. Department of the Environment, Transport and the Regions, London, 39–53
15. Brookfield JFY (1996) A simple new method for estimating null allele frequency from heterozygote deficiency. *Mol Ecol* 5: 453–455
16. Excoffier L, Smouse P, Quattro J (1992) Analysis of molecular variance inferred from metric distances among DNA haplotypes: application to human mitochondrial DNA data. *Genetics* 131: 479–491
17. Kresovich S, Szewc-McFadden AK, Bliek SM, McFerson JR (1995) Abundance and characterisation of simple-sequence repeats (SSRs) isolated from a size-fractionated genomic library of *Brassica napus* L. (rapeseed). *Theor Appl Genet* 91: 206–211
18. Szewc-McFadden AK, Kresovich S, Bliek SM, Mitchell SE, McFerson JR (1996) Identification of polymorphic, conserved simple sequence repeats (SSRs) in cultivated *Brassica* species. *Theor Appl Genet* 93: 534–538
19. Slatkin M (1995) A measure of population subdivision based on microsatellite allele frequencies. *Genetics* 139: 457–462
20. Hall TS (1995) The Molecular Population Genetics of Self-Incompatibility in a Natural Population of *Brassica oleracea*. Ph.D. Thesis, University of Birmingham
21. Weber JL, Wong C (1993) Mutation of human short tandem repeats. *Hum Mol Genet* 2: 1123–1128
22. Evans KA, Allen-Williams LJ (1992) Electroantennogram responses of the cabbage seed weevil, *Ceutorhyncus assimilis*, to oilseed rape, *Brassica napus*, volatiles. *J Chem Ecol* 18: 1641–1659

23. Bartlet E, Blight MM, Hick AJ, Williams IH (1993) Responses of the cabbage seed weevil (*Ceutorhyncus assimilis*) to the odor of oilseed rape (*Brassica napus*) and to some volatile isothiocyanates. *Entomol Exp Appl* 68: 295–302
24. Blight MM, Pickett JA, Wadhams LJ, Woodcock CM (1995) Antennal perception of oilseed rape, *Brassica napus* (*Brassicaeae*), volatiles by the cabbage seed weevil *Ceutorhyncus assimilis* (Coleoptera, Curculionidae)
25. Smart LE, Blight MM (1997) Field discrimination of oilseed rape, *Brassica napus* volatiles by cabbage seed weevil, *Ceutorhyncus assimilis*. *J Chem Ecol* 23: 2555–2567
26. Moyes CL, Raybould AF (1998) Herbivory by the cabbage seed weevil (*Ceutorhyncus assimilis*) in natural populations of *Brassica oleracea*. *Acta Hort*; *in press*
27. Warman EA (1998) Gene flow in wild cabbage. *In*: AJ Gray, F Amijee, CJ Gliddon (eds): *Environmental Impact of Genetically Modified Crops*. Department of the Environment, Transport and the Regions, London, 39–53
28. Gómez JM, Zamora R (1994) Top-down effects in a tritrophic system: parasitoids enhance plant fitness. *Ecology* 75: 1023–1030
29. Kling J (1996) Could transgenic supercrops one day breed superweeds? *Science* 274: 180–181
30. American Institute of Biological Sciences (1995) *Transgenic Virus-Resistant Plants and New Plant Varieties*. American Institute of Biological Sciences, Washington
31. Cooper JI (1997) Might transgenes conferring virus resistance harm the environment? *In*: D McLean, PM Waterhouse, G Evans, MJ Gibbs (eds): *Commercialisation of Transgenic Plants: Risk, Benefit and Trade Considerations*. Bureau of Resource Sciences, Kingston ACT, Australia, 115–124
32. Edwards ML, Cooper JI (1985) Plant virus detection using a new form of indirect ELISA. *J Virol Meth* 11: 309–319
33. Skotniki ML, Mackenzie AM, Ding SW, Mo JQ, Gibbs AJ (1993) RNA hybrid mismatch polymorphisms in Australian populations of turnip yellow mosaic virus. *Arch Virol* 132: 83–99
34. Paul VH, Rawlinson CJ (1992) *Diseases and Pests of Oilseed Rape*. Verlag Th. Mann, Gelsenkirchen-Buer
35. Bartsch D, Schmidt M, Pohl-Orf M, Haag C, Schuphan I (1996) Competitiveness of transgenic sugar beet resistant to beet necrotic yellow vein virus snd potential impact on wild beet populations. *Mol Ecol* 5: 199–205
36. Bergelson J (1994) Changes in fecundity do not predict invasiveness: a model study of transgenic plants. *Ecology* 75: 249–252
37. Dinsh Kumar SP, Whitham S, Choi D, Hehl R, Corr C, Baker B (1995) Transposon-tagging of tobacco mosaic virus resistance gene N – its possible role in the TMV-N-mediated signal transduction pathway. *Proc Natl Acad Sci USA* 92: 4175–4180
38. Broer I (1996) Stress inactivation of foreign genes in transgenic plants. *Field Crop Res* 45: 19–25

Methods for Risk Assessment of Transgenic Plants
III. Ecological risks and propspects of transgenic plants
K. Ammann, Y. Jacot, V. Simonsen and G. Kjellsson (eds)

A multisite-cooperative research programme on risk assessment of transgenic crops

Xavier Reboud, Jacques Gasquez and Henri Darmency

Laboratoire de Malherbologie, INRA, BV 1540, F-21034, Dijon, France

Introduction

Genetically modified plants are now being commercialised in several countries as regulatory authorities consider that the balance of risk *versus* benefit is beneficial. However, numerous questions remain unanswered, especially the impact of these plants when used over large areas and under a range of variable environmental conditions. Some issues need to be re-evaluated [1, 2]. Risk/safety analysis, as well as prospects of transgenic crops depend on the scale which is to be considered. Extrapolation of methods, and laboratory and greenhouse results, to large-scale farmers' fields, may provide useful preliminary data, but is not a sound approach to the study of the consequences of the commercial release of transgenic crops. Risk/safety analysis involves hazard identification and risk assessment. Hazards are scale-dependent and need to be tested on an appropriate scale. Change to a region's flora, for instance, is not a matter for a greenhouse test, but must be studied at the regional level. Risks associated with identified hazards depend on local conditions, e.g. non-proportionality of pollen spread with source size, regional variation of crop management practices, interaction between crops, genotype variability of populations of wild relatives, etc. A main concern is to estimate the value of predictability for agriculture and environment of results collected on different scales. In order to answer this question, several institutes (Association Générale des Producteurs de Maïs AGPM, Centre Technique Interprofessionnel des Oléagineux Métropolitains CETIOM, Institut National de la Recherche Agronomique INRA, Institut Technique de la Betterave ITB, Institut Technique des Cérélaes et des Fourrages ITCF), seed producers (KWS, Novartis) and agrochemical companies (Agrevo, Monsanto, Rhone-Poulenc) jointly designed an experiment on an agricultural scale. This was a multisite testing of several gene constructs with different crops rotating in farmers' fields over several years, representing a true field situation.

Material and methods

Experimental design: Three sites were chosen in Northern, Eastern and Southern France representing different sides of the country, climatic conditions and typical regional cropping systems. Transgenic varieties were introduced into the crop rotation in the place of normal sugar beet, rapeseed and maize. Winter wheat and fallow were also used. The five adjacent plots were 1 ha each. Experiments were conducted as closely as possible to that of the usual farming practice of each region. Transgenes were herbicide resistance against glyphosate, glufosinate and bromoxynil in rapeseed (three varieties), glyphosate and glufosinate in sugar beet and maize (two varieties each). Bt transformed maize was also used. The main objectives were to quantify the ease of use for the farmer, agronomic performance with regional average, and assess potential problems. Well characterised markers allowed the study of pollen flow around the crops, gene flow between varieties, occurrence of multi-resistant volunteers, behaviour and control of volunteers, and hybridisation between crops and wild relatives.

Pollen dispersal

Dispersal of pollen outside the field was assayed by the introduction of male-sterile plants, at various distances from the field plots, to biologically trap pollen of the corresponding crop. At maturity, seed production was recorded and seeds were collected and tested for resistance by screening a portion of the sampled seed with herbicide. Resistance within the progeny of these plants provided evidence of pollination by the crop.

Cross-pollination between varieties

Cross-pollination was studied by taking seed samples prior to harvest at different positions within the field. Treatment of volunteers (rapeseed) was also used as an easy way to estimate cross-pollination within the field. For a mother plant resistant to herbicide A, the progeny was screened with the other herbicide(s). A second treatment with A was also conducted to remove potential seed pollution and thus ascertain the expected double resistance.

Volunteers

Crop rotation allowed determination of the occurrence of transgenic volunteers in other crops as well as studies on their natural ability to grow, flower and produce progeny. Multi-resistance was also tested using non-destructive Elisa tests on volunteers. The most interesting results can be expected after the second year of the experiment.

Potential gene escape to wild relatives

All wild relatives of rapeseed and sugar beet occurring within a 1000 m diameter circle in the centre of the field were mapped. The potential hybridisation between a crop and its wild relatives was estimated by sampling all stands of wild relatives previously observed. At maturity, a proportion of the seeds were collected and screened in the greenhouse for resistance. As samples sizes were relatively small, the probability of observing hybridisation was low and one or two orders of magnitude below what may have been tested in the laboratory. Remaining seeds were left in the habitat in order to not interfere too much with local flora and potential weed infestation.

Results

Rapeseed

Pollen dispersal and effective isolation of the experiment from rapeseed grown by other farmers was checked on plots of male-sterile plants at different positions. Cross-pollination occurred between varieties within the experiment. With each pollen source of approximately 1/3 ha, cross-pollination decreased with distance, giving 2.5% at 1 m, 0.18% at 22 m and slightly less than 0.01% hybrids at 65 m. At 127 m no cross-pollination was observed so that the rate of cross-pollination at that distance could not exceed 2/10 000 seeds with 95% confidence. However, an isolated natural volunteer rapeseed growing in 1996 in a fallow at 150 m from the nearest corner of the field gave two seedlings resistant to Bromoxynil and 1 resistant to Glyphosate. This suggested that the rate of cross-pollination is highly dependent on the size of the pollen source and the area of recipient plants. Volunteers were observed in maize but as they were winter rape, they could not have a complete life cycle. They may also occur in subsequent years in wheat and sugar beet. As for the wild relatives of rapeseed, no hybrid was found. Hybridisation rates would be lower than those for which we could test according to the numbers of emerged seedlings of each species tested, i.e. 0.01% for *Sinapis alba,* 0.03% for *Sinapis arvensis* (with 95% confidence).

Sugar beet

Sugar beet is a biannual crop grown for its root, so very few flowering plants were expected. One variety did not flower, while the other had 24 plants which flowered over half a hectare in 1996. The analysis of seed production of male sterile plots showed that 3% of the amount of pollen at 0 m of pollen was still available 190 m from the field. Volunteers, observed in both fallow ground and winter wheat, could flower and set seed. In order to test the potential for crosses between sugar beet and its weedy relative, several plots of annual weedy plants were added along one field border. In addition, in one location, about 200 volunteers occurred naturally in the fallow plot. The progenies of those plants gave a total of 40 000 seedlings of which 23 were resistant. According to the number of plants flowering within the field trial, the plants bearing resistant pollen only represented about 5% of field flowering plants. As such, the cross-pollination observed may be underestimated by one order of magnitude.

Maize

Asynchrony of flowering between varieties limited the study of gene flow. Seed production on emasculated plants at 150 m dropped to about 10% of the seed amount observed on fully pollinated plants.

Conclusion

Although our agronomic approach of risk assessment on a large scale seems to be well fitted, it has overall limited value and is time consuming! Trying to both monitor and carry out experiments in true agricultural situations is a compromise. For instance, the accuracy of measurements made for gene flow and hybridisation depends on the number of seeds and plants tested, but only a small proportion of the seeds produced by the wild plants were studied. Of course, most of the seed produced must remain in the field if we want the weed community to evolve under normal field conditions. Therefore, some issues could not be correctly evaluated in such a study. Additional simplified approaches are required to reach a significant value of predictability. For instance, the spontaneous production of hybrids by wild relatives of rape has been shown to occur at different rates according to the wild species [3–5]. The question is not only the frequency of hybridisation, but rather what may occur after several years. There might be invasion by descendants of hybrids, change of weed flora due to the use of the new herbicides, or new problems with volunteers etc. Leaving the field system evolving also enables us to check secondary effects on non-target organisms such as the adjacent wild flora and non-target insects (in the case of the Bt resistant maize), which is much harder to test experimentally.

References

1. Darmency H (1996) Potential disadvantages of herbicide-resistant crops in weed resistance management. *In*: *Second Int Weed Control Congress, Copenhagen*, Dept Weed Control, Flakkegjerg, 427–433
2. Cooper JI, Raybould AF (1997) Transgenes for stress resistance: consequences for weed evolution. *In*: *Brighton Crop Protec Conf Weeds*, BCPC, Farnham, 265–272
3. Mikkelsen TR, Andersen B, Jorgensen RB (1996) The risk of crop transgene spread. *Nature* 380: 31
4. Lefol E, Danielou V, Darmency H (1996) Predicting hybridisation between transgenic oilseed rape and wild mustard. *Field Crop Res* 45: 153–161
5. Lefol E, Fleury A, Darmency H (1996) Gene dispersal from transgenic crops. Hybridisation between oilseed rape and the hoary mustard. *Sex Plant Reprod* 9: 189–196

Monitoring the environmental impact of transgenic sugar beet *Beta vulgaris* subspec. *vulgaris altissima* Döll – are we able to ask the right questions?

Matthias Pohl-Orf[1], Ulrike Brand[2], Ingolf Schuphan[1] and Detlef Bartsch[3]

[1]*Department of Biology V, Ecology, Ecochemistry and Ecotoxicology, RWTH-Aachen University of Technology, Worringerweg 1, D-52056 Aachen, Germany*
[2]*Institute for Developmental Biology, University Cologne, Gyrhofstr. 17, D-50923 Köln, German*
[3]*present address: University of California, Riverside, Department of Botany and Plant Sciences, Riverside, CA 92521, USA*

Introduction

With the release and the commercialization of transgenic plants the spread of genetically modified phenotypes in the environment seems certain. The remaining question is merely how long this process will take and what effects it will cause. The fundamental problem when talking about risk assessment is that we lack the necessary extensive knowledge about the ecology of the modified species and the function of the transferred genes. Only this knowledge makes it possible to look for the really relevant topics and to ask the right questions for an adequate risk assessment. To address ecologically important features like outcrossing [1], competitiveness [2], gene flow [3] or survival through the winter the following investigations were carried out. To make the aims of the investigation clearer it is useful to differentiate between two kinds of monitoring. Following Maas [4], there is on the one hand specific monitoring which examines possible cause-effect relationships, and on the other hand a general monitoring which focuses on investigations for example in natural populations without using the transgenic organism.

Our specific monitoring was primarily carried out in single organism tests with the transgenic plants themselves to check the ecologically relevant winter survival. Our general monitoring focuses on the virus infestation level in the potentially influenced habitats and on the genetic structure of the populations of the wild relatives. A definition of the actual state of population dynamics using RAPD-PCR was necessary for an evaluation of possible future changes in natural populations due to transgene introgression. These data give information about whether a specific trait is able to cause a competitive advantage in a particular habitat and thus help to assess the potential risk of outcrossing and establishment of transgenic traits in wild beet habitats. Only precise knowledge about such transgene/environment interactions enables us to identify priorities for research.

Material and methods

Plant material

The breeding lines, cultivars and transgenic varieties were made available to us by KWS/PLANTA, Einbeck, Germany. The transgenic sugar beets we worked with were carrying the additional transgenic sequences of the c-DNA of the coat protein of beet necrotic yellow vein virus (BNYVV) [5], the nptII gene [6] as a resistance marker against kanamycin and the bar-gene [7] mediating resistance against the herbicide BASTA/LIBERTY with its active agent glufosinate-ammonium. Some wild varieties were from FAL, Braunschweig, Germany, while others were from own collections.

Virus infestation

12 to 16 days after germination, the seedlings were placed in soil containing BNYVV. Thirty-two plants of each treatment were watered twice a week with sea salt water at two different concentrations. The first treatment was 1% salt, the second 0.5% and the third was tap water as a control. After cultivation for 95 to 106 days, the roots were harvested and checked for the presence of BNYVV [8–10] with a specific antibody test (ELISA, enzyme linked immunosorbent assay) [11, 12].

Survival of sugar beet in the winter

Between 1994 and 1997, field tests were performed at two sites, one with virus infestation (and pre-inoculation) in Mainz, Germany, the other at a virus free control site in Aachen, Germany. From 1994 to 1996 the tests were conducted with conventional plants done at different sites in Germany (Braunschweig, Dresden, Aachen, Köln, Stuttgart) and at the Dutch coast near Breskens.

Molecular analysis of population genetics

The RAPD-PCR (randomly amplified polymorphic DNA-Polymerase chain reaction) was based on Lorenz et al. [13], Uphoff and Wricke [14, 15] and Eagen and Goldman [16], optimized for the specific requirements like used enzymes or DNA-template. (PCR-conditions: 10 mM buffer, 1.5 mM $MgCl_2$, 0.2 mM dNTPs, 1u taq, 0.001% gelatin, 0.5 µg primer, approx. 100 ng template, added to 50 µl with water; amplification program: denaturation 30 sec, 94 °C, annealing 1 min, 35 °C, synthesis 2 min, 72 °C, 40 cycles, DNA-isolation: SDS-method, Taq-polymerase: EUROGENTEC Goldstar, Primer: MWG-Biotech).

Results

Virus infection under different soil conditions

The typical extinction value of an infected plant was about 2.5 after 60 min of reaction time. Transgenic plants expressing the virus coat protein and thus responding to the test showed values as a positive control of approximately 0.5. The different salt concentrations in the water caused a decrease of infection with an increase of salt content in the water. As shown in Figure 1, the decrease is significant except in wild beet population #1, which showed the highest infection under 0.5% salt solution watering.

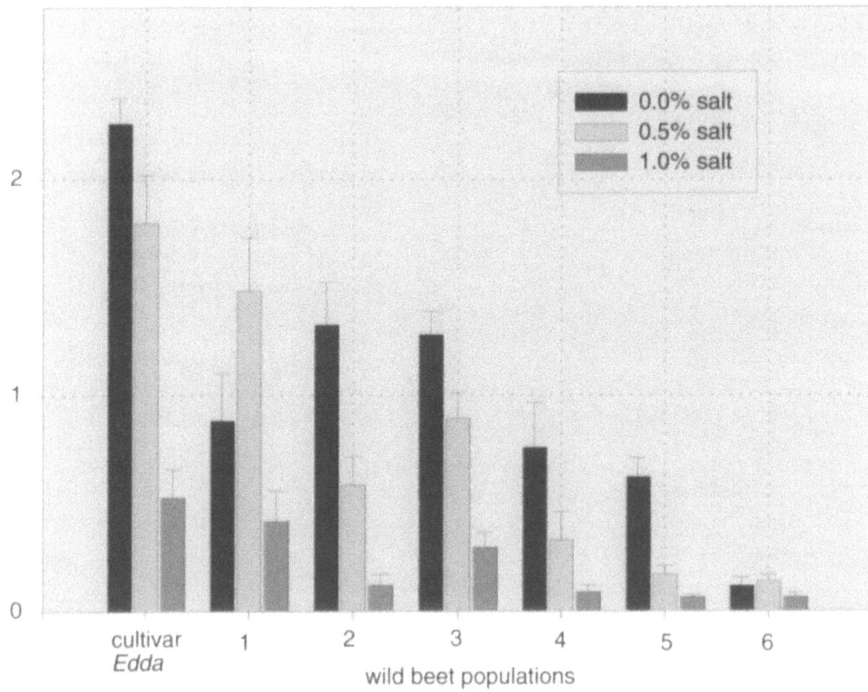

Figure 1. Infection of different wild populations of *Beta vulgaris* L. with BNYVV depending on salt concentration in the soil (mean of 32 plants with standard error). The six different wild beet populations originated from the Adriatic coast in Italy.

The different wild variants showed a different range of extinction values. One of the populations even proved to be completely tolerant against virus infection. In this case the extinction never exceeded a value of 0.3 in the ELISA. Especially in the case of the cultivar Edda, morphological changes due to salt irrigation were noticed. The plants developed succulent characteristics such as thick leaves with a strong cuticle and more compact growth thus causing morphological similarity to their wild relatives.

Survival of transgenic and conventional sugar beet

As shown in Figure 2, correlation was found between survival rate and temperature in the trials when survival rate was plotted against cold sum. The independent variable cold sum is defined as the sum of every daily negative average temperature at 2 m above ground over the whole winter. Summing only average temperatures below –4 °C gave better correlations and this parameter was chosen as independent variable. The logistic regression with the formula $f_{(x)} = (100)/(1 + (x/c)^b)$ and the parameters b = 4.213 and c = –32.461 had a correlation coefficient of r = 0.931.

Significant differences between the genotypes at the same test site were not detected and so the different genotypes were combined and compared within the different sites. In the winter

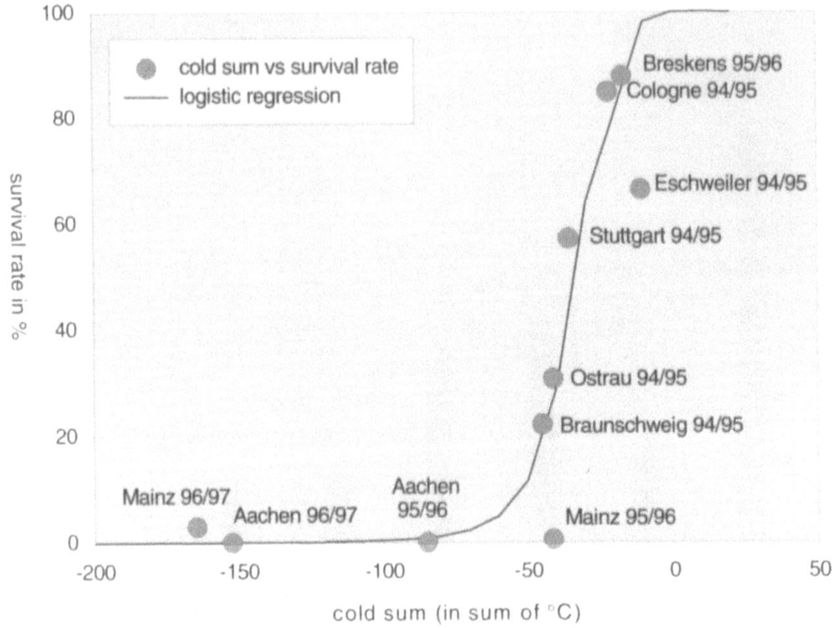

Figure 2. Survival rates of *Beta vulgaris* L. at different sites in different test periods (mean over all genotypes). Transgenic plants were used in Mainz and Aachen, Germany. The tests in 1996/97 were carried out with hybrids between sugar beet and Swiss chard.

1994/1995 the temperature was moderate and the survival rates were high. At the field site in Cologne more than 90% of the beets survived. The winters of 1995/1996 and 1996/1997 were much harder and survival rates were so low that nearly all plants died (Fig. 2). Only in Breskens at the Dutch border were the temperatures milder and here survival rates were high (90%).

Characteristic RAPD marker for wild and cultivated beets

Four primers showed suitable banding patterns and produced 24 useful bands. To find particularly reproducible fragments, only amplification products of a specific size (depending on the primer, between 400 and 2000 bp) were evaluated. Five markers were found that occurred only in wild beets and two were found to be characteristic for cultivars.

Discussion

Specific monitoring

Even in very harsh winters, survival of sugar beet in North Europe is possible. Contrary to earlier investigations [17] according to which sugar beet will die at temperatures below –5 °C, plants survived minimum temperatures of –10 °C and less. The comparison of different genotypes in a species, especially in the case of the Swiss chard hybrids, may be matter of discussion. Swiss chard, being a leaf crop, is not like the other beet varieties selected towards big beetroots. Small beetroots have a relatively higher content of osmotic compounds and less water causing a potentially higher frost resistance [18].

General Monitoring

Our investigations show that the infection of wild beets with BNYVV is low in natural, mesohaline habitats. This is very important due to the ecological relevance of transgenic rhizomania resistance. The occurrence of *Beta vulgaris* ssp. *maritima* is limited to a small area along the coasts where a moderate salt concentration reveals. This could be the reason why BNYVV has not been found in wild beets grown in such natural habitats. In addition to this, wild beets are genetically less susceptible to rhizomania [8] and so wild beets carrying the transgenic virus resistance would not behave much differently from their unaffected relatives in salt free habitats where an infection could occur. Whether the decreasing infection at higher salt concentration directly depends on the virus or on the disturbance of the vector *Polymyxa betae* KESKIN has not been clarified, but the important aspect is that a transgenic virus resistance is of minor ecological importance due to the lack of selection pressure. However, the fact that a tolerance against BNYVV in wild beet is found seems to be a hint towards an co-evolutionary process and thus the temporary occurrence of BNYVV in wild beet habitats.

The RAPD-PCR resulted in several specific marker bands which are characteristic either for wild beets or for cultivars. These markers are useful tools to investigate the interaction between wild and cultivated beets and the great heterogeneity within the species *Beta vulgaris* [19], and support the hypothesis of creation of the weed beets by hybridization during seed production near wild beet populations, as in Italy and France [20, 21].

The present investigation gives an impression of how many aspects have to be observed in an extensive monitoring program. It should be clear that all modified traits of a transgenic plant have to be checked towards their possible ecological consequences based on the qualities of the new gene. Moreover, the conditions in the habitats of the wild relatives are of great importance to evaluate the relevance of the selective advantage caused by the new traits.

It seems to be a never ending story – research will only be done if effects of a change can be imagined. Who, for example, thought of CO_2 production and global change when introducing automobiles instead of horse carriages?

The question remains

Do we know enough about plant ecology to be sure we have considered all the possible implications of releasing transgenic plants? Thus, the reason for many of the "reassuring" results could be that we have not looked for the right topics and have not asked the right questions.

Acknowledgement
We want thank all who made the investigations possible due to their cooperation especially: KWS/PLANT, Einbeck, Dr. L. Frese, FAL-Braunschweig, Dr.Posselt and Mr. Stelz University Stuttgart-Hohenheim, MPI-Züchtungsforschung, Köln, Dr. E. Biancardi, Instituto Sperimentale per le culturi, Rovigo. The work was funded by the German Ministry of Education and Science (No. 0310532).

References

1. Bartsch D, Pohl-Orf M (1996) Ecological aspects of transgenic sugar beet: Transfer and expression of herbicide resistance in hybrids with wild beets. *Euphytica* 91: 55–58
2. Bartsch D, Schmidt M, Pohl-Orf M, Haag C, Schuphan I (1996) Competitiveness of transgenic sugar beet resistant to beet necrotic yellow vein virus and potential impact on wild beet populations. *Molec Ecol* 5: 199–205
3. Bartsch D, Schmidt M (1997) Influence of sugar beet breeding on populations of *Beta vulgaris* ssp. maritima in Italy. *J Veg Sci* 8: 81–84
4. Maas D (1996) Möglichkeiten des Biomonitorings bei der Langzeitbeobachtung transgener Organismen. – *In*: Langzeitmonitoring von Umwelteffekten transgener Organismen. *UBA-Texte* 58/96: 47–60
5. Meulewaeter F, Soetaert P, Emmelo Van J (1989) Structural analysis of the coat protein gene in different BNYVV Isolates. *Medelingen Faculteit Landbouwwetenschap Riijksuniversiteit Gent* 54(2): 465–468
6. Beck E, Ludwig G, Auerswald EA, Reiss B, Schaller H (1982) Nucleotid sequence and exact localization of the neomycin phosphotransferase gene from transposon Tn5. *Gene* 19: 327–336
7. Thompson CJ, Mova NR, Tizard R, Crameri R, Davies JE, Lauwereys M, Botterman J (1987) Characterization of the herbicide-resistance gene bar from Streptomyces hygroscopicus. *EMBO J* 3: 2723–2730
8. Whitney E, D (1989) Identification, distribution, and testing for resistance to rhizomania. *Beta* maritima. *Plant Dis* 73(4): 287–290
9. Alderlieste M F J And Van Eeuwijk FA (1992) Assessment of concentrations of beet necrotic yellow vein virus by enzyme-linked immunosorbent assay. *J Virol Meth* 37: 163–176
10. Geyl L, Garcia Heriz M, Valentin P, Hehn A, Merdinoglu D (1995) Identification and characterisation of resistance to rhizomania in an ecotype of *Beta vulgaris* ssp. maritima. *Plant Pathol* 44: 819–828
11. König R, Burgermeister W, Leseman D-E (1987) Methods for Detection and identification of Beet Necrotic Yellow Vein Virus. *Proc. 50th Winter Congress, I.I.R.B.*, Brussels, 17–22
12. Kaufmann A, König R, Lesemann D-E (1992) Tissue print-immunoblotting reveals an uneven distribution of beet necrotic yellow vein and beet soil-born viruses in sugarbeets. *Arch Virol* 126: 329–335
13. Lorenz M, Weihe A, Börner T (1994) DNA fragments of organellar origin in random amplified polymorphic DNA (RAPD) patterns of sugar beet (*Beta vulgaris* L.). *Theor Appl Genet* 88: 775–779
14. Uphoff H, Wricke G (1992) Random Amplified Polymorphic DNA (RAPD) Markers ın Sugar Beet (*Beta vulgaris* L.): Mapping the Genes for Nematode Resistance and Hypocotyl Color. *Plant Breeding* 109: 168–171
15. Uphoff H, Wricke G (1995) A genetic map of sugar beet (*Beta vulgaris*) based on RAPD markers. *Plant Breeding* 114: 355–357
16. Eagen K A, Goldman I L (1996) Assessment of RAPD marker frequencies over cycles of recurrent selection for pigmnet concentrtion and percent solids in red beet (*Beta vulgaris* L.). *Mol Breeding* 2: 107–115
17. Anonymous (1993) Report results from frost resistance trials with sugar beet (*Beta vulgaris* L.) transformed with glyphosate resistance genes. Reports to the National Agency for Environmental Protection. Denmark
18. Barocka KH (1985) Zucker- und Futterrüben, *In*: *Lehrbuch der Pflanzenzüchtung landwirtschaftlicher Kulturformen, Bd. 2, Spezieller Teil.* Paul Parey Verlag, Berlin-Hamburg, 245–287
19. Jung C, Pillen K, Frese L, Fähr S, Melchinger AE (1993) Phylogenetic relationship between cultivated and wild species of the genus *Beta* revealed by DNA "fingerprinting". *Theor Appl Genet* 86: 449–457
20. Boudry P, Mörchen M, Saumitou-Laprade P, Vernet P, Van Dijk H (1993) The origin and evolution of weed beets: consequences for the breeding and release of herbicide-resistant transgenic sugar-beets. *Theor Appl Genet* 87: 471–478
21. Santoni S, Berville A (1992) Evidence for gene exchange between sugar beet (*Beta vulgaris* L.) and wild beets: consequences for transgenic sugar beets. *Plant Mol Biol* 20: 578–580

Methods for Risk Assessment of Transgenic Plants
III. Ecological risks and propspects of transgenic plants
K. Ammann, Y. Jacot, V. Simonsen and G. Kjellsson (eds)

Discussion session 1: Ecological effects of transgenes

Alan Raybould
My work aims at answering the question whether herbivore have any effect in natural viral populations, if they drive the population dynamics. If the herbivore had a big effect in a viral population, maybe we should look at the effects of insect resistance transgene. If not, then perhaps it might be less of a problem.
The viruses used in my experiments might be cross-protected.

Jim White
Cross-protected viruses are naturally occurring. And this is true for many viruses and their host plants. And the one reason is all these viruses are aphid transmitted viruses.

Ian Cooper
Viruses do tend to cross-protect against closely related species.
There is some tendency for the naturally occurring viruses to be mild in their effects. However they are not mild in their effects on test-plants when taking a test elsewhere.

Jim White
I want to underline the fact that if you take the virus strain out of a natural situation and place it under artificial condition, you can get different kinds of symptoms. One of the very important things is that the vector has to come to the plant and for the aphid-transmitted ones, they need immediatly to feed on the new host plant. And the symptoms depend on how the plants are growing in that area, the season, and probably the culture conditions. And there is change from year to year.

Alan Raybould
Time is probably important and has probably a log-effect.

Tom Nickson
It was also pointed out that it is not wise to talk about insect resistance as a very general thing. We have to study what the transgene is really doing.
Considering the monitoring, it was pointed out that we should consider monitoring over a period of years for instance when we look at plant survival. It is essential because climatic conditions vary so much.

Matthias Pohl-Orf
The only thing we can do is to spread the field sites over a big area, so you have very different climatic conditions.

Phil Dale
Just a question about the oilseed rape the glyphosphate and bromoxylene: I think that the studies on herbicide resistant plants give us useful data, but I wonder how and to what extent we could have predicted this kind of data from what we already know about gene flow? We need to know much more about ecological consequences if combining different herbicide tolerances in agriculture is a reality. What are the crucial questions? We need to commercialize and grow transgenic plants before we can answer the grander question about the ecological and the agronomic impact. So I think the question is: what are the really crucial bits of information we need before we move forward to approve all three herbicide tolerance genes in oilseed rape?

Les Levidow
An important point is: What more do we need to know before commercialization of transgenic plants? This question should be also examined at the regulatory level.
In the early days of these risk assessment discussions it was sometimes said that herbicide tolerance genes, virus resistance genes and so on, are wide spread in nature and in some crops. And therefore we had a basis for predicting the effect of such transgenes into crops which did not have those particular genes before. The presentation of session 1 and this discussion takes us very far away from that origin of the familiarity criteria. Instead, we have

a set of interesting questions about the unfamiliarity of virus resistance, herbicide tolerance, etc., for example What is the extent of virus infection? If the virus infection is low or even high in particular wild relatives, then why? On what is it dependent? I want to pose a general conceptional question for us to think about what is given at the risk assessment research starting from a presumption of unfamiliarity. How can these different questions about unfamiliarity be related? If we rightly ask questions about the unfamiliarity of how plants are or are not infected by viruses or resistant to viruses, then methodically, conceptionally, how can the different unfamiliar aspects be related? And therefore how can these different risk assessment projects be related.

Phil Dale

Familiarity can be defined at very different levels. It is realistic to compare familiarity at a phenotypic level where we may have varieties or whatever it is that carry at a phenotypic level resistance to a particular virus. But at a genetic level, a functional level, the mechanism achieving the phenotype may be different. And in that case we may not have complete familiarity. You know there are aspects of unfamiliarity and it raises questions about recombination at codon switching, and you know of possible recombination between the invading virus and the homology sequences on the virus transformed into the plants. So we think there are different levels of familiarity and unfamiliarity; we have to draw on what information we can and build that into the risk assessment process.

Tom Nickson

Let me point to the example of Canada, where genetically modified oilseed rape has been approved now for 3 years in production. Transgenic rape represent around half of the production of oilseed rape in 1998. Canadian authorities looked into the risk assessment of out-crossing and the obvious probability that it will occur to *Brassia rapa*. But they also accepted a risk management regarding gene escape at a regulatory level, and at an industrial agricultural level. This is something that is not being balanced in the discussion. Where does the risk-management come in? And along those lines, we have the opportunity to create something that is new and very powerful: the area of monitoring. Monitoring has to be for regulatory reasons to ensure the safety at the production level and this is something that industry can accept. Monitoring has to be accepted by the industry. But there also has to be a joint collaboration between academia, industry and government to develop a more fundamental knowledge of the ecology in agriculture. This collaboration is extremely important: The benefits will be evident in developing sustainable agriculture project scenarios. We must continue to ask all these very detailed questions, but for the purpose of developing sustainable agriculture projects not for the purposes of blessing technology.

Session 2: Modelling in risk assessment

Methods for Risk Assessment of Transgenic Plants
III. Ecological risks and propspects of transgenic plants
K. Ammann, Y. Jacot, V. Simonsen and G. Kjellsson (eds)

The role of modelling in risk assessment for the release of genetically engineered plants

Glynis D. Gidding

Institute of Biological Sciences, Sir George Stapledon Building, Aberystwyth, Ceredigion, SY23 3DD, UK

Introduction

Models can help us to understand, and make predictions about, the behaviour of biological systems. This includes those which will govern the fate of transgenes in the environment. The suitability of various models is considered for predicting and monitoring the potential for invasiveness [1, 2], or causing outbreeding depression [3, 4] or genetic assimilation [5]. The influences of spatial and temporal heterogeneity and stochasticity are discussed in relation to their effects on predictive modelling. The importance of modelling in the design of risk assessment experiments [6], for summarizing experimental data [7] and in monitoring transgene spread after release [6], is demonstrated.

Statistical models of invasions

Considerable statistical modelling has been conducted in an attempt to correlate invasiveness with biological, genetic and/ or environmental traits (e.g. [6, 8–11]). From this it has been deduced that vegetative and reproductive traits are not good predictors of invasiveness *per se* [11–14]. Whether a plant is annual or perennial, or if it is an inbreeder, outbreeder or asexual, is not an indicator of its potential invasiveness [10, 12] Genetic characteristics – such as polyploidy and levels of variability or heterozygosity – are also poor predictors of colonising ability [8]. Differences between plants which succeed and fail are often apparently trivial [12] and *may be determined by just a few genes* [11, 13, 15].

It has been estimated that about 10% of organisms deliberately or accidentally introduced into the environment have feral populations [12]. Of these about 10% establish permanent populations, with about 10% of these becoming pests (the "tens rule" [11, 13]). Williamson and Fitter [16] considered exceptions to this, one example of which was British edible crop plants. They found that because crops are strongly selected to grow well in the region where they are cultivated almost all non-native crop plants were at least casual, i.e. with temporary feral populations (95%). They concluded that the tens rule does not apply if there has been selection to counteract it. This will be true in the case of genetically modified crop plants.

This, coupled with the severe consequences of invaders which do have adverse ecological or agricultural consequences [2, 17–19], warrants the use of regulation and monitoring as an insurance policy. It is clearly inappropriate to base predictions about potential risks solely according to the qualitative properties of transgenics.

Mathematical models of invasions

Two of many possible models are used to illustrate what could be required for any particular model of escaped transgenics. The first is a reaction-diffusion model adapted, by Hastings [20], from a spatial Lotka model suggested by van den Bosch and colleagues [21]. The second was developed by Crawley [22] to describe the population biology of invaders.

The Hastings model

In the model presented by Hastings [20] the birth rate at time t and location x is given as

$$b(t,x) = \int_{space}\int_{age} b(t-a,\psi)L(a)\times m(a)D(a,|x-\psi||alive)\ da\ d\psi$$

For a plant $b(t, x)$ is the probability of a seedling at time t and location x. $L(a)$ is the probability of surviving to age a and $m(a)$ the fecundity at age a. Movement is described by a dispersal kernel which can be interpreted as the probability that a seedling germinated a time units ago at location ψ has now produced a seedling at location x. Discrete time and/or space versions of reaction-diffusion models may be required for plants [20]. Models of this sort lead to travelling-wave population densities of constant velocity [20, 23].

The Crawley model

Crawley [22] considers the rate of change of population size of species i over time:

$$\frac{dN_i}{dt} = r_iN_i - f_1(\psi,N_i) - f_2\left[\sum_{j=1}^{i-1} N_j, N_i\right] - f_3\left[\sum_{k=1}^{i-1} N_k, N_i, P\right] - f_4(M,N_i) + X$$

where r_i is the rate of increase per individual under ideal conditions (intrinsic rate of increase). The minimum condition for population increase is, therefore, $r_i > 0$. The functions $f_1 - f_4$ characterise what are often termed the density dependent functions.

f_1 represents the way in which the resource supply rate ψ affects the abundance of species i. ψ has different meanings in the "Crawley" and "Hastings" models and it should be realised that there is no connection between them. f_2 describes the effects of interference competition where species $j = 1 \cdots (i-)$ are "fiercer" than the invader. Individuals of one species reduce the fitness of others in ways not directly related to resource capture, e.g. by allelopathy or alteration of the physical environment. f_3 represents the influence of natural enemies (P), and is related to their preference for the invader in comparison to other prey species, where $k = 1 \cdots (i-)$ are species they would prefer to prey on. With regard to plants the enemies are diseases and herbivores. f_4 describes negative effects due to a limited supply of mutualists, e.g. rhizobia, mycorrhizae and pollinators. Indirect effects of mutualists may also be accounted for in $f_1 - f_3$, for example predators of natural enemies will reduce P in f_3.

X is what Crawley describes as the "mystery" ingredient essentially representing immigration into the population, including recruitment from protected refuges, and is a way of introducing heterogeneity into the model.

A comparison of the Hastings and Crawley models

The Hastings model deals with a particular space and time and the probability that it will become occupied by the species of interest whereas the Crawley model deals with the rate of change of the number of individuals of a particular population over time. Both models deal with the occupancy of space, by a species of interest, over time.

In the first model occupancy here and now depends on the probability that individuals born *elsewhere* and previously have *produced offspring* that *survive* here and now. In the second model population change depends on the number of *surviving offspring per individual* and immigration. For both the main parameters are fecundity, survivorship and dispersal. Crawley's model explicitly specifies the influences on these as exploitation competition, interference competition, natural enemies, mutualists and the presence of refuges.

It is possible to incorporate stochasticity into either model by using Monte Carlo computer simulations which draw values from probability density functions instead of using fixed parameters. Also some choices of function for the dispersal kernel in the Hastings model would necessarily involve different population attributes over space, for example one involving the pollen dispersal models of Giddings et al. [24].

The importance of gaps

Recruitment from seed in an existing community depends on there being suitable gaps in the vegetation where germination can occur. Small changes in gap density, due to grazing animals, for example, can result in large changes in recruitment [25–27]. Models which describe this [e.g. 25–27] are related to the Hastings model presented above. If b(t,x) was at sometime a gap, then what it is now depends on the probability that what was in existence then dispersed seeds into the gap which germinated into seedlings and survived to now.

Not all invaders reproduce sexually. Japanese knotweed (*Reynoutria japonica*), for example, has spread around the UK despite only female plants being present. Some transgenic crops may reproduce asexually, in fact male sterility may be a feature introduced into some, such as forage grasses, in order to prevent cross-pollination with indigenous plants. Simple models, some such version of a Richardson model [20, 28], can be used to investigate the likely spread of vegetative propagules accidentally released outside the crop. Such models are simplified discrete versions of the Hastings model in which the neighbourhood set defines dispersal of clonal material rather than seeds.

Neighbourhood models

Neighbourhood models were developed specifically for studying the population dynamics of sessile organisms [34, 35]. They are currently very popular for practical applications (e.g.

[29–34, 38, 39]), particularly cellular automata models. They have been used to model clonal growth (e.g. [36, 37]), gap colonisation by annuals (e.g. [40, 44]), competition between annuals (e.g. [29, 42]), between annuals and perennials (e.g. [25]), and between perennials (e.g. [43]), succession (e.g. [23, 44–46]) and weed spread (e.g. [47–50]). Interactions are usually assumed to be local and dispersal limited. There may be many variables concerned with an explicit description of the various underlying mechanisms of density dependence. It may also be possible to model the effects of farming and cultural practices. This makes such models more realistic but less tractable than more strategic models. For this reason they are usually implemented as computer simulations (e.g. [36, 37]).

Such "realistic" models can be treated as "experiments" which can be done (relatively) quickly and on a large scale compared with field experiments. Parameters can be varied systematically, experimental simulations replicated, and the outcome analysed statistically. That is not to suggest models replace field experiments, which are required to provide parameter estimates for the models in the first place. They can however demonstrate the likely spread of transgenes in particular conditions and the possible effects of stochasticity on predictions.

A model for transgenic oilseed rape

First order difference equations relating population sizes of succeeding generations can be used to model the likely invasiveness of annuals. Crawley et al. [7] used the simple density independent model $N_{t+1} = \lambda N_t$ for investigating the potential invasiveness of oilseed rape, where λ is the finite rate of population change. λ can be estimated from the ratio of seedlings sampled in $t + 1$ to seedlings sampled in t or defined as a function of the probabilities of germination, survival to flowering and surviving offspring produced per individual.

When $\lambda = 1$, $N_{t+1} = N_t$ and the population is at equilibrium. The invasion criteria is, therefore, $\lambda > 1$. At equilibrium the change in population over time is zero. Recall from Crawley's model that, for a population in ideal conditions modelled in continuous time, $\frac{dN_t}{dt} = rN_t$ and $r = 0$ at equilibrium. If for any particular non-ideal conditions $\frac{dN_t}{dt} = r' N_t$ then $N_t = N_0 e^{r't}$ and the multiplication rate between two consecutive times is $e^{r'}$. Hence it is seen that the relationship between λ and r is $e^{r'} = \lambda$, or $r' = \ln\lambda$.

The rape model is simply a discrete time version in which the mechanisms underlying the departure from "ideal" are hidden. Functions $f_1 - f_4$ and X in Crawley's model define the difference between r and r'. Although the rape model does not include explicit density dependence the experimental data was collected from 12 different habitats to which different treatments were applied, at three sites in the UK and over 3 years. Interspecific plant competition was the main determinant of λ which was only greater than one when the seeds were distributed in a cultivated competition free environment from which other vegetation had been removed. When surrounding perennial vegetation reinvaded the cultivated plots λ again dropped to below one highlighting the importance of competition free gaps for recruitment into natural habitats. Thus what appears to be a density independent model is, in reality, density dependent by virtue of the manner in which the data has been collected.

Introducing stochasticity into these models is easily done by using a probability density function for λ. Log transformation converts such multiplicative models into additive ones. This is useful for parameter estimation from data, and because the probability density function of

$\log(\lambda)$ is arithmetic rather than geometric [51] (invasion criteria $\log(\lambda) > 0$). Some problems of estimating λ from experimental data are considered below.

Model modifications for perennials

Perennial populations can be structured according to age or stage. Possible stages might be: (1) germination and seedling development when survivorship is low and fecundity zero; (2) pre-sexual maturity when survivorship has increased but fecundity is still zero; (3) sexual maturity when survivorship is high and fecundity above zero and (4) senescence when survivorship and fecundity decline to zero. The perennial equivalent of the difference equation presented for oilseed rape involves using age or stage structured population projection matrices (M) and population structure vectors (v_t). The information required for these models is $s_{t,t+1}$, the survivorship from time t to $t + 1$ and f_t the fecundity at time t, for each age or stage. A value for λ can be determined from the dominant eigenvalue of the matrix (for a clear explanation of why and how to do this see chapter 5 of Gillman and Hails [52]).

Mechanistic approaches to defining rates of population change

In some cases there will be advantages in quantifying the rate of population change by formulating some or all of the Crawley functions $f_1 - f_4$ and X, for example, when transgenes enhance traits involved in competition by conferring resistance to pests and pathogens. Isolation from pests and pathogens appears to be an important factor influencing the success of invaders [11, 53]. Induced resistance to pests and pathogens is therefore worthy of particular attention with regard to risk assessment [1, 54, 56]. Andow [1] suggested that Tilman's resource acquisition models [56–59] could be used for predicting the effect of transgenic resistance to pests and pathogens such as insects, viruses, bacteria and fungi.

Tilman's models predict the dynamics of plant communities based on the way plants acquire and use resources [56–59]. The relative growth rate (*RGR*) of a population of plant species i is estimated as per unit change in biomass (B), i.e. the difference between growth from resource use and biomass loss from any source. Hence

$$\frac{dB_i}{B_i dt} = f_i(R) - m_i$$

where t is time, $f_i(R)$ is a function giving the rate of biomass increase from using resource R, and m_i is biomass loss, here independent of plant biomass and the resource. Each species i has its own associated equation.

The rate of resource change is the rate of resource renewal $y(R)$ minus resource removal by all species in the community:

$$\frac{dR}{dt} = y(R) - \sum_i [Q_i B_i f_i(R)]_i$$

where Q_i is the nutrient content per unit biomass of plant species i and the summation is over all species in the community.

R_i^* is the concentration of a particular resource that species i must have to persist without changing its biomass, i.e. the minimum resource requirement for persistence. If more than one species are limited by the same resource then the species with the lowest R_i^* will increase the fastest and *eventually* replace the others, *all other things being equal*. Transgenics with low R_i^* are therefore more likely to be invasive than those with a high minimum resource requirement. Transgenes which lower R_i^*, for example by providing resistance to pests which consume plant material, could turn an otherwise non-invasive plant (either crop or crop-native hybrid) into one which could alter community structure. Transgenics with high R_i^* might persist in local patches due to spatial heterogeneity, or have relatively high rates of colonisation and low mortality [60–61]. They will not replace species with low R_i^* and are therefore unlikely to substantially change community structure[1] (unless, for example, there is evolution which lowers R_i^*). Not all transgenic resistance will increase biomass accumulation, e.g. if pests attack grains or spoil fruits, when effects on fecundity may be important.

Genetic considerations

Transgenic-native hybrids might become invasive, displacing other species in the environment [62–65]. If transgenes bestow some measurable advantage on their host then modelling the fate of individuals in the hybrid population is possible using neighbourhood models such as cellular automaton. This requires assigning genotypes to individuals in the model and determining differences in phenotype from experimental data.

There are two other possible situations where genetic introgression of transgenes into natural populations of conspecifics may be a problem: (1) where persistent immigration of transgenes causes outbreeding depression [66–68], and (2) where genetic assimilation causes the extinction of the original species [68]. In the first transgenes bestow some measurable *disadvantage* on their host, which, when there is a continued immigration of transgenes, could cause a decline in the native population. The second case might be particularly important in the case of inter-specific hybridisation where the composition of the population is important, although if the fitness of hybrids is greater than that of the parents the abundance of the population could also change. Both can be modelled by tracking individuals.

What can models tell us about experimental design and sampling effort?

Having good parameter estimates is important for modelling. One crucial question in relation to designing experiments to provide these is: "in how many different sites and years should demographic parameters be measured". We might further ask what magnitude of error we would expect if we reduce the size and duration of experiments to below this "ideal".

Kareiva et al. [6] conducted a particularly illuminating analysis of the data presented by Crawley et al. [7]. First they calculated an average rate of population change over the full range of environments and years. They calculated 100 geometric mean rates of population change for each site by drawing and replacing multiplication rates at random from the original data, and then took arithmetic means of these. They calculated an average canola multiplication rate of 0.11 with 95% confidence intervals ranging from 0.01 to 0.46. Assuming that this represented "reality" for the whole set of data they repeated the process for different subsets of the data, simulating the results that would have been obtained from different numbers of sites and years (i.e. from smaller experiments). The error due to reduced sampling was presented as the percentage difference from the overall mean. Data from 1 or 2 years observations were from several hundred percent to about 100% different from the overall data respectively, regardless of the number of sites. Data from 3 years obviously became more accurate with increasing numbers of sites.

This large variation in plant success between years is not unusual [49] and indicates that experimental risk assessments will require several years of data. Anything less is likely to greatly reduce the predictive power of any models derived from it. This is particularly so when the average multiplication rate is close to one, i.e. when it is not obvious whether the transgenic is destined for extinction, persistence or spread. In these cases long-term monitoring of transgenics may be relevant. Such monitoring could be continued to a time when subsets of the data accurately predict the whole, i.e. the data is apparently as variable as it is likely to be. The analysis of Kareiva et al. [6] shows a useful method for assessing the reliability of data obtained for risk assessment experiments, and hence the usefulness of models derived from it.

Predictable prediction problems: heterogeneity and stochasticity

Predictive modelling can only be successful if it deals with situations which are predominantly deterministic. Particularly when populations are small, and in the case of long-term predictions, stochastic processes might override deterministic ones [20, 35, 69, 70]. Escapes from transgenic crops occurring at low frequencies will be vulnerable to stochastic events. Then the occurrence of repeated escapes from widely grown crops may be important in determining whether transgenics establish feral populations [64]. Stochasticity can be incorporated into models by taking random draws from a specified probability distribution function [52, 66]. It may be demographic, environmental [71] or genetic.

In patchily distributed populations where there are repeated extinction-recolonisation events, for example when there are repeated escapes from commercially grown crops, gene frequencies will be determined by the number and genotype of founders [69, 70]. Then genetic drift [72] could have a major influence on the fate of the transgenes [69, 70]. Introgression [66, 73], epistasis [73–75], pleiotropy [76–80], spontaneous chromosome doubling or the introduction of transgenic gene controlling sequences [54] could all affect the fitness of hybrids containing transgenes, and thus their persistence in the environment [80, 81].

Monitoring models

An analysis of 90 different invasive weeds in the northwestern United States found that the initial extent and rate of increase was not a good predictor of their eventual distribution [6]. This should caution us against using deterministic models to make long-term predictions. Demographic parameters estimated from a transgenic crop are likely to alter with environmental stochasticity and the *evolution* of feral populations. If a transgene can be maintained in casual or naturalised populations then sooner or later evolution, or the chance occurrence of favourable circumstances, might promote its hosts to invasion status. This situation provides a strong argument for long-term monitoring and modelling of the demography of feral populations harboring transgenes of potential ecological consequence.

Summary and conclusions

1. Statistical modelling of biological invasions shows it is inappropriate to base predictions about the potential risks of transgenics solely according to their qualitative properties. Modelling can help to make some predictions concerning the risk of invasion, genetic assimilation and outbreeding depression.
2. Neighbourhood models can be used to simulate the population dynamics of escaped transgenics and their hybrids with native plants. Survivorship, fecundity and dispersal can be specified according to some version of the Hastings model, while the output, in terms of population growth, can be formulated in the style of the Crawley model.
3. Quantifying survivorship, fecundity and dispersal has to be done experimentally. The mechanisms that regulate these are exploitation competition, interference competition, natural enemies, mutualists and the presence of refuges. Farming and cultural practices may also be important. These may or may not be explicitly taken into account when measuring and modelling survivorship, fecundity and dispersal.
4. The reliability of data obtained for risk assessment experiments, and hence the usefulness of models derived from it, can be tested using the method of Kareiva et al. [6].
5. When populations are small, and in the case of long-term prediction, stochastic processes might override deterministic ones, making reliable predictions difficult or impossible. Evolution might promote benign escapees to invasion status – a situation that could be impossible to predict in advance. Risk assessors need to be aware of such difficulties – and exercise caution – when considering model predictions of the risks of transgenics (or any other predictions, for that matter). Particularly as "off the shelf" models become available, users must be aware of the underlying assumptions and limitations.
6. The possible unreliability of small data sets and influence of stochasticity provide a strong argument for long-term monitoring and modelling of the demography of feral populations harboring transgenes of potential ecological consequence.

Models, being quantitative, can appear deceptively precise. We must, therefore, be careful not to overestimate their predictive capabilities. Nevertheless they are of undoubted value and their use will increase our knowledge of both transgenics and the release environment.

References

1. Andow DA (1994) Community response to transgenic plant release: using mathematical theory to predict effects of transgenic plants. *Molec Ecol* 3: 65–70
2. Paoletti MG, Pimentel D (1996) Status and prospects of genetic engineering in agriculture and the environment. *Minerva Biotechnol* 8: 195–207
3. Svensson L (1988) Inbreeding, crossing and variation in stamen number in *Scleranthus annuus* (Caryophyllaceae), a selfing annual. *Evolutionary Trends in Plants* 2: 31–37
4. Svensson L (1990) Distance-dependence regulation of stamen number in crosses of *Scleranthus annuus* (Caryophyllaceae) from a discontinuous population. *Amer J Bot* 77: 889–896
5. Simberloff D (1988) The contribution of population and community biology to conservation science. *Annu Rev Ecol Syst* 19: 473–511
6. Kareiva P, Parker IM, Pascual M (1996) Can we use experiments and models ın predicting the invasiveness of genetically engineered organisms. *Ecology* 77: 1651–1675

7. Crawley MJ, Hails RS, Rees M, Kohn D, Buxton J (1993) Ecology of transgenic oilseed rape in natural habitats. *Nature* 363: 620–3
8. Gray AJ (1986) Do invaders have definable genetic characteristics? *Phil Trans R Soc Lond B* 314: 655–674
9. Williamson M, Brown KC (1986) The analysis and modelling of British invasions. *Phil Trans R Soc Lond B* 314: 505–522
10. Perrins J, Williamson M, Fitter A (1992) A survey of differing views of weed classification: implications for regulation of introductions. *Biol Conserv* 60: 47–56
11. Williamson M (1994) Community response to transgenic plant release: predictions from British experience of invasive plants and feral crop plants. *Molec Ecol* 3: 75–79
12. Williamson M (1993) Invaders, weeds and risks from GMOs. *Experientia* 49: 219–224
13. Williamson M (1996) Can the risks from transgenic crop plants be estimated. *Trends Biotech* 14: 449–450
14. Lawton JH (1990) Biological control of plants a review of generalisations, rules and principles using insects as agents. *In*: C Bassett, LA Whitehouse, AJ Zabkiewicz (eds): *Alternatives to the chemical control of weeds.* FRI Bulletin 155. New Zealand Ministry of Forestry, Wellington, New Zealand, 3–17
15. Williamson M (1992) Environmental risks from the release of genetically modified organisms (GMOs) – the need for molecular ecology. *Molec Ecol* 1: 3–8
16. Williamson M, Fitter A (1996) The varying success of invaders. *Ecology* 1661–1666
17. OTA [USCongress Office of Technology Assessment] (1993) Harmful non-indigenous species in the United States. US Government Printing Office, Washington DC, USA
18. Bergelson J, Newman JA, Floresroux EM (1993) Rates of weed spread in spatially heterogeneous environments. *Ecology* 74: 999–1011
19. Higgins SI, Richardson DM, Cowling RM (1996) Modelling invasive plant spread: the role of plant-environment interactions and model structure. *Ecology* 77: 2043–2054
20. Hastings A (1996) Models of spatial spread: is the theory complete? *Ecology* 77: 1675–1679
21. van den Bosch F, Metz JAJ, Diekmann O (1990) The velocity of spatial population expansion. *J Math Biol* 28: 529–565
22. Crawley MJ (1986) The population biology of invaders. *Phil Trans R Soc Lond B* 314: 711–731
23. Cz(r(n T, Bartha S (1989) The effect of spatial pattern on community dynamics – a comparison of simulated and field data. *Vegetatio* 83: 229–239
24. Giddings GD, Sackville Hamilton NR, Hayward MD (1997) The release of genetically modified grasses. Part 2: the influence of wind direction on pollen dispersal. *Theor Appl Genet* 94: 1007–1014
25. Crawley MJ, May RM (1987) Population dynamics and plant community structure: competition between annuals and perennials. *J Theor Biol* 125: 475–89
26. Silvertown J, Smith B (1989) Germination and population structure of spear thistle *Cirsium vulgare*in relation to experimentally controlled sheep grazing. *Oecologia* 81: 369–73
27. Klinkhamer PGL, De Jong TJ (1989) A deterministic model to study the importance of density dependence in for regulation and the outcome of intra-specific competition in populations of sparse plants. *Acta Bot Neer* 38: 57–65
28. Durett R, Levin S (1994) Stochastic spatial models: a user's guide to ecological applications. *Phil Trans R Soc Lond B* 343: 329–350
29. Weiner J, Conte PT (1981) Dispersal and neighborhood effects in an annual plant competition model. *Ecol Model* 13: 131–147
30. Mithen R, Harper JL, Weiner J (1984) Growth and mortality of individual plants as a function of "available area". *Oecologia* 57: 57–60
31. Pacala SW (1986) Neighbourhood models of plant-population dynamics. 2. multi-species models of annuals. *Theor Popl. Biol* 29: 262–292
32. Pacala SW, Silander JA, Jr, (1985) Neighborhood models of plant-population dynamics. 1. single-species models of annuals. *Amer Naturalist* 125: 385–411
33. Pacala SW, Silander JA, Jr, (1985) Neighborhood predictors of plant performance. *Oecologia* 66: 256–263
34. Pacala SW, Silander JA, Jr, (1990) Tests of neighbourhood models in field communities of two annual weed species. *Ecol Monogr* 60: 113–134
35. Cz(r(n T, Bartha S (1992) Spatiotemporal dynamic models of plant populations and communities. *TREE* 7: 38–42
36. Barkham JP, Hance CE (1982) Population dynamics of the wild daffodil (*Narcissus pseudonarcissus*). III. Implications of a computer model of 1000 years of population change. *J Ecol* 70: 323–344
37. Inghe O (1989) Genet and ramet survivorship under different mortality regimes – a cellular automata model. *J Theor Biol* 138: 257–270
38. Hara T (1988) Dynamics of size structure in plant-populations. *Trends Ecol Evol* 3: 129–133
39. Slatkin M, Anderson DJ (1984) A model of competition for space. *Ecology* 65: 1840–1845
40. Hobbs RJ, Hobbs VJ (1987) Gophers and grassland: a model of vegetation response to patchy soil disturbance. *Vegetatio* 69: 141–146
41. Perry JN, Gonzalez-Andajur JL (1993) Dispersal in a metapopulation neighbourhood model of an annual plant with a seed bank. *J Ecol* 81: 453–63
42. Czárán T (1989) Coexistence of competing populations along an environmental gradient: a simulation study. *Coenoses* 4: 113–120
43. Silvertown J, Holtier S, Johnson J, Dale P (1992) Cellular automata models of interspecific competition for space. *J Ecol* 80: 527–34
44. Hogeweg P, Hesper B, Schail CP van Beeftink WG (1985) Patterns in vegetation succession, an ecomorpho-

logical study. *In*: J. White (ed.): *The population structure of vegetation*. W. Junk, Dordrecht, 637–666
45. van Tongeren O, Prentice IC (1986) A spatial simulation model for vegetation dynamics. *Vegetatio* 65: 163–173
46. Colansanti RL, Grime JP (1993) Resource dynamics and vegetation processes: a deterministic model using two-dimensional cellular automata. *Funct Ecol* 7: 169–76
47. Auld BA, Coote BG (1980) A model of spreading plant population. *Oikos* 34: 287–292
48. Auld BA, Coote BG (1981) Prediction of *Nassella trichotoma* (Gramineae) invasion of pasture in south-east Australia. *Prot Ecol* 3: 271–277
49. Auld BA, Coote BG (1990) INVADE: Towards the simulation of plant spread. *Agr Ecosyst Environ* 30: 121–128
50. Auld BA, Vere DT, Coote BG (1982) Evaluation of control policies for the grassland weed, *Nassella trichotoma* (Gramineae) in south-east Australia. *Prot Ecol* 4: 331–338
51. Foley P (1994) Predicting extinction times from environmental stochasticity and carrying capacity. *Conserv Biol* 8: 124–37
52. Gillman M, Hails R (1997) *An introduction to ecological modelling. Putting practice into theory*. Blackwell Science, UK. 202 pp
53. Lawton JH, Macgarvin M (1986) Interaction between bracken and its insect herbivores. *Phil Trans R Soc Edinburgh B* 86 125–131
54. Tiedje JM, Colwell RK, Grossman YL, Hodson RE, Lenski RE, Mack RN, Regal PJ (1989) The planned introduction of genetically engineered organisms: ecological consideration and recommendations. *Ecology* 70: 298–315
55. Schmitt J, Linder CR (1994) Will escaped transgenes lead to ecological release? *Molec Ecol* 3: 71–74
56. Tilman D (1982) *Resource competition and community structure*. Princeton University Press, Princeton, New Jersey, US
57. Tilman D (1988) *Plant strategies and the dynamics and structure of plant communities*. Princeton University Press, Princeton, New Jersey, US
58. Tilman D (1990) Mechanisms of plant competition for nutrients: the elements of a predictive theory of competition. *In*: JB Grace, GD Tilman (eds): *Perspectives on plant competition*. Academic Press, San Diego, 117–141
59. Tilman D (1994) Competition and biodiversity in spatially structured habitats. *Ecology* 75: 2–16
60. Hamilton WD, May RM (1977) Dispersal in stable habitats. *Nature* 269: 578–581
61. Yodzis P (1978) *Competition for space and ecological communities*. Springer-Verlag, New York
62. Raybould AF, Gray AJ (1993) Genetically modified crops and hybridization with wild relatives: a UK perspective. *J Appl Ecol* 30: 199–219
63. Giddings GD, Sackville Hamilton NR, Hayward MD (1997) The release of genetically modified grasses. Part 1: Pollen dispersal to traps in *Lolium perenne*. *Theor Appl Genet* 94: 1000–1006
64. Templeton AR (1986) Coadaption and outbreeding depression. *In*: ME Soule (ed.): *Conservation biology: the science of scarcity and diversity.* Sinauer, Sunderland, UK, 105–116
65. Ellstrand NC, Hoffman CA (1990) Hybridization as an avenue of escape for engineered genes. Strategies for risk reduction. *BioScience* 40: 438–441
66. Simberloff D (1988) The proximate causes of extinction. *In*: DM Raup, D. Jablonski. (eds): *Patterns and processes in the history of life.* Berlin, Springer-Verlag. 259–76
67. Ellstrand NC, Hoffman CA (1990) Hybridization as an avenue of escape for engineered genes. Strategies for risk reduction. *BioScience* 40: 438–441
68. Ellstrand NC (1992) Gene flow by pollen: implications for plant conservation genetics. *Oikos* 63: 77–86
69. Ellstrand NC (1992) Gene flow by pollen: implications for plant conservation genetics. *Oikos* 63: 77–86
70. Templeton AR (1986) Coadaption and outbreeding depression. *In*: ME Soule (ed.): *Conservation biology: the science of scarcity and diversity.* Sinauer, Sunderland, UK, 105–116
71. May RM (1973) *Stability and complexity in model ecosystems*. Princeton University Press. Princeton, New Jersey
72. Wright S (1970) Random drift and the shifting balance theory of evolution. *In*: K. Kojima (ed.): *Mathematical topics in population genetics*. Springer-Verlag, Berlin, 1–31
73. Colwell RK, Norse EA, Pimentel D, Sharples FE, Simberloff D (1985) Genetic engineering in agriculture. *Science* 229: 111–112
74. Lenski RE (1988) Experimental studies of pleiotropy and epistasis in *Escherichia coli*. I. Variation in competitive fitness among mutants resistant to virus T4. *Evolution* 42: 425–432
75. Lenski RE (1988) Experimental studies of pleiotropy and epistasis in *Escherichia coli*. II. Compensation for maladaptive effects associated with resistant to virus T4. *Evolution* 42: 433–440
76. Guivarch A, Spena A, Noin M, Besnard C, Chriqui D (1996) The pleiotropic effects induced by the rolc gene in transgenic plants are caused by expression restricted to protophloem and companion cells. *Transgenic Res* 5: 3–11
77. Olesinski AA, Lucas WJ, Esra G, Shmuel W (1995) Pleiotropic effects of tabacco-mosaic-virus movement protein on carbon metabolism in transgenic tobacco plants. *Planta* 197: 118–126
78. Überlacker B, Klinge B, Weer W (1996) Ectopic expression of the maize homeobox genes ZmHox1a or ZmHox1b causes pleiotropic alterations in the vegetative and floral development of transgenic tabacco. *Plant Cell* 8: 349–362
79. Sawahel WA (1994) Transgenic plants: performance, release and containment. *World J Microbiol Biotechnol* 10: 139–144

80. Giddings G, Mytton L, Griffiths M, McCarthy A, Morgan C, Skøt L (1997) A secondary effect of transformation in *Rhizobium leguminosarum* transgenic for *Bacillus thuringiensis* subspecies *tenebrionis* δ-endotoxin (cryIIIA) genes. *Theor Appl Genet* 95: 1062–1068
81. Jones S (1995) An evolutionists view of genetic engineering. *In*: H Bernhard, C Cookson (eds): *Genethics. Debating issues and ethics in genetic engineering.* Ciba-Geigy Limited

Methods for Risk Assessment of Transgenic Plants
III. Ecological risks and propspects of transgenic plants
K. Ammann, Y. Jacot, V. Simonsen and G. Kjellsson (eds)

Modelling the spread of disease resistance gene in natural plant populations

Christian Damgaard

Department of Terrestrial Ecology, National Environmental Research Institute, Silkeborg, Denmark

The biotechnological development of disease resistance is going to be increasingly important for future transgenic plants. Therefore, it is appropriate to investigate the effect of a potential spread of resistance genes to other populations or species on the natural environment. Here modelling becomes an important tool in estimating probabilities and evaluation of the risks of the spread of a resistance gene.

The modelling of the potential spread of disease resistance genes in natural plant populations requires that characteristic host - pathogen interactions are taken into account.

The infection process of a typical airborne plant pathogenic fungi depends on the density of host plants [1, 2].

The pathogen population size (census and effective) is typically large and the selection pressure on the pathogen population is high.

The genetic basis of resistance and virulence varies from a gene-for-gene system of host recognition [3–5] to quantitative resistance (partial resistance) and induced resistance.

Difference in life history characteristics (e.g. whereas a typical plant relies on a "seed bank strategy" to ensure presence in a local subpopulation, pathogens frequently goes locally extinct and rely on long distance dispersal of spores [6]).

Here, I will summarise the results of two complementary models of evolution in the gene-for-gene system: A metapopulation model with local extinction and recolonisation, and a spatial model of the spread of a resistance gene in a local population. Both models will be published in detail elsewhere.

A well known plant resistance mechanism against fungal diseases is host recognition of the pathogen typically followed by a local hypersensitive cell death response [7]. The host is assumed to recognise the pathogen by a fungal gene product which either is involved in the general maintenance of the fungus or the actual infection process. The genetics of the recognition process is most easily explained by a single host locus and a corresponding pathogen locus. The host locus has two functional alleles, a usually dominant resistance allele and a susceptible allele. The pathogen locus, likewise, has two alleles, a usually dominant avirulence allele and a virulence allele. A host carrying a resistance allele is resistant to the avirulent pathogen, whereas a host without a resistance allele is susceptible to both pathogen phenotypes [5].

When natural host-pathogen populations are investigated, a considerable amount of polymorphism in the gene-for-gene system has been observed [2, 8]. This is surprising because the selection pressure in favour of the resistance and especially the virulence allele is high, when there is variation in the system. Normally, the observed variation is explained by "cost of resistance" and "cost of virulence" [9]. However, when investigated, it has been difficult to measure any significant costs of either resistance or virulence, possibly because it is difficult to measure selection on a single gene in different genetic backgrounds [2, 5]. Another explanation may be that at the metapopulation level the host and pathogen populations never reach equilibrium, but instead the local populations continually are experiencing mutation and immigration of especially new pathogen genotypes.

Here, the dynamics of the gene-for-gene-system is examined in a metapopulation of an annual selfing host and a haploid pathogen subpopulations. Migration of both hosts and pathogens occur among the subpopulations either according to an island migration model or a stepping-stone migration model. Extinction and recolonization is modelled according to the characteristic life-history differences between plants and pathogens, thus if a plant subpopulation goes extinct the plant population is recolonized from the seedbank whereas the pathogen is recolonized by migration. If it is assumed that the probability of extinction is a function of the disease level in the subpopulation, then it is possible to maintain variation in the gene-for-gene system at the metapopulation level at a significantly higher level than by mutation-selection balance.

Additionally, the fixation probabilities and mean fixation times of a resistance allele in an annual partial selfing local host population which is infected by a monomorphic pathogen population is examined. In the model, there is a number of pathogen generations for each plant generation. The mortality of the plants is a function of whether or not the plants are infected and when they were infected. The infection and plant birth processes are assumed to be partly dependent on the state of the neighbours. One conclusion of the modelling is that the breeding system of the host plant has a large effect on the probability that a resistance allele is fixated in the population. If only the female fecundity is affected by the disease, i.e. not the pollen production, then the probabilities of fixation in some cases are higher for outcrossing populations compared to selfing populations. Another conclusion is that the probability of fixation depends on the spatial structure of the host population and the infection process.

Combining results from models with local neighbour dependent birth and death processes, and extinction and recolonization processes at the level of the metapopulation will lead to an understanding of which effects are most important for predicting the potential spread of a resistance gene on a local and a regional scale, and it will be possible to suggest which parameters should be estimated in order to predict possible ecological scenarios.

We, together with Risø Research Institution and Copenhagen University, are starting a project to investigate the potential spread of a disease resistance gene in *Brassica napus* into a *Brassica campestris* population. The project will focus on the probability that a neutrally or positively selected gene will cross the hybridisation barrier by estimating the competitive ability of the different back-cross generations in field experiments. The following spread of a resistance allele into the *B. campestris* population from a low frequency due to hybridisation, as well as the effect on the pathogen population, will be evaluated/predicted using models as discussed above.

References

1. Antonovics J (1994) The interplay of numerical and gene-frequency dynamics in host-pathogen systems. *In*: LA Real (ed.): *Ecological genetics*. Princeton University Press, Princeton, 129–145
2. Burdon JJ (1987) *Disease and plant population biology*. Cambridge University Press, Cambridge
3. Flor HH (1942) Inheritance of pathogenicity in *Melampsora lini*. *Phytopathology* 32: 653–668
4. Flor HH (1955) Host parasite interaction in flax rust. – Its genetics and implications. *Phytopathology* 45: 680–685

5. Simms EL (1996) The evolutionary genetics of plant-pathogen systems. *BioScience* 46: 136–145
6. Thompson K (1992) The functional ecology of seed banks. *In*: M Fenner (ed.): *Seeds the ecology of regeneration in plant communities.* Redwood Press, Melksham, 231–258
7. Goodman RN, Novacky AJ (1994) *The hypersensitive reaction in plants to pathogens. A resistance phenomenon*, APS Press, St. Paul, Minnesota
8. Parker MA (1985) Local population differentiation for compatibility in an annual legume and its host-specific fungal pathogen. *Evolution* 39: 713–723
9. Leonard KJ (1994) Stability of equilibria in a gene-for-gene coevolution model of host-parasite interactions. *Phytopathology* 84: 70–77

Methods for Risk Assessment of Transgenic Plants
III. Ecological risks and propspects of transgenic plants
K. Ammann, Y. Jacot, V. Simonsen and G. Kjellsson (eds)

The wave of advancement of introduced genes in natural plant populations

Jarle Tufto

ICAPB, University of Edinburgh, King's Buildings, West Mains Road, Edinburgh EH9 3JT, UK

Introduction

In plants, gene flow is limited to relatively short distances, implying that dynamic models which aim to predict the fate of genes introduced by genetic engineering need to take into account the geographic location of the genes present in the population at any particular point in time. The frequency p of some alleles is then a function not only of time t but also of the geographic location x. Several different formulations of such spatially explicit models are given in the literature. Here, I give a brief overview of these different models and some of the predictions which may be relevant when discussing the long-term spread of introduced genes. I also present a numerical solution for the resulting dynamics in one particular situation and discuss some implications of these results for the design of monitoring programs.

The model

We consider a population distributed continuously in one single dimension x. Local population density is constant and large, such that local genetic drift can be ignored, and generations are discrete. Let $p_t(x)$ denote the frequency of the transgene in generation t at geographic location x. We now assume that gene displacements, through seed and pollen dispersal, around every plant in the population follow some probability distribution $f(x)$. The frequency of the transgene also changes locally due to selection. With no dominance this gene frequency change is equal to $sp_t(x)[1 - p_t(x)]$. With these assumptions, the frequency of the transgene in generation $t + 1$ at each geographic location x becomes

$$p_{t+1}(x) = \int f(y-x)\, p_t(y)\mathrm{d}y + sp_t(x)[1-p_t(x)]. \qquad (1)$$

This equation defines the dynamics of a spatially explicit model with selection and dispersal acting every generation on local gene frequencies. The frequency of the transgene as a function of geographic location and time, is completely determined by (1) and the initial state of the population, that is $p_t(x)$ at $t = 0$. It should be noted that this model can also be applied to a population distributed in two dimensions, as long as there is no dependency on the second dimension y in the initial state of $p_t(x,y)$, and as long as the selection coefficient s is independent of y.

Analytic results

It is clear that if a selectively favoured gene is released at some geographic location, it will at first increase in frequency locally, and, at the same time, disperse into new parts of the population. After some initial generations, a steady state moving wave front will be formed. By doing certain approximations, Fisher and Kolmogorov [1, 2] independently showed that this wave front will move with a constant speed of

$$\nu = \sqrt{2s\sigma} \qquad (2)$$

Note that the dispersal probability distribution $f(x)$ in the first term of the right-hand side of (1) only appear through a single parameter σ^2 in (2), defined as the variance of the gene displacements, that is, $\sigma^2 = \int x^2 f(x)\mathrm{d}x$. This parameter, the dispersal variance (or its square root, the dispersal standard deviation) is the relevant quantity to estimate, and should not be confused with another frequently used measure of dispersal, the mean dispersal distance $\int |x| f(x)\mathrm{d}x$, which in general has no simple relationship to σ^2. The dispersal variance can be estimated from different forms of data, either directly by using pollen traps [3] or genetic markers [4, 5], or indirectly from the amount of genetic differentiation in subdivided populations [6]. In general, unless a random set of, say, pollen grains can be followed, assumptions need to be made about the distributional form of $f(x)$. Some estimates of σ and some other measures of dispersal are given in Table 1.

If we use $\sigma = 40$ m, assume a selection coefficient of, say, $s = 0.1$, and also rely on Fisher's approximation, then this implies that the transgene will move with a speed of only $v = 18$ m/generation. For organisms with long generation times, such as many species of trees, the distance by which the wave front will move each year may become very small. Norway spruce, for example, has a generation time of about 20 years, which means that we should expect the wave front to move only about 1 m/year. Although these calculations can only be taken as a rough approximation, they may still give us an idea about the relevant temporal and geographic scale of the problem.

Numerical analysis

The analytic results presented above refer to the asymptotic behaviour of the wave front, that is, the speed at which the wave front moves when stationarity has been attained many generations after the introduction. During the first few initial generations, however, the behaviour of

Table 1. Some estimates of σ and other measures of dispersal

Species	Method	Estimate	Reference
Meadow fescue (*Festuca pratensis*)	Genetic markers	30.0 m	[5]
Sea beet (*Beta vulgaris maritima*)	Indirect	74.6 m	[6]
Norway spruce (*Picea* abias)		30.6 m*	[7]
Pinus cembriodes		16.8 m*	[7]

* Assuming an exponential dispersal distribution.

the wave front may be quite different. The gene frequency dynamics will still be completely determined by (1) and the initial state of the system, but needs to be analysed using numerical techniques. This can be done by slightly modifying the model, by assuming that the population is distributed in many small discrete subunits, instead of assuming that space is continuous. Using discrete coordinates, model (1) can be rewritten to

$$p_{t+1,i} = \sum_j m_{ij} p_{t,j} + s p_{t,i}\left[1 - p_{t,i}\right], \quad (3)$$

where m_{ij}, determined by $f(x)$, represents the probability of dispersal to subunit i from some geographic location in subunit j. The dynamics of the model can then be simulated by setting all $p_{0,i}$ to some initial value, and then iterating (3) for the desired number of generations.

Two such simulations are shown in Figure 1, the first with selective advantage ($s = +0.1$), and the second with selective disadvantage ($s = -0.1$). In both cases, dispersal of genes follow a double exponential (Laplace) distribution with a standard deviation of 40 m. The simulated populations consists of two regions; one transgenic "cultivated" part at $x < 0$ in which the frequency of the transgene is maintained at $p_{t,i} = 1$, say by replantation of the crop every year, and a natural part at $x > 0$ in which the trangene, in the initial generation, is assumed to be absent. The resulting gene frequency clines at every fifth generation are shown.

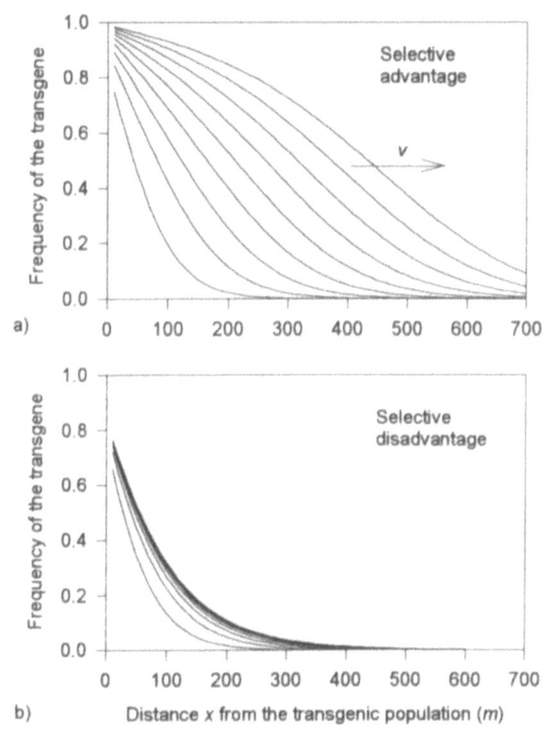

Figure 1. The frequency of the transgene as a function of distance from a permanent, transgenic population, with (a) selective advantage, and (b) selective disadvantage. The gene frequency cline at every 5th generation is shown. Further details are given in the main text.

There are some interesting points to notice. Even when there is selection against the transgene several dispersal standard deviations still become established in the natural part of the population. After about 30 generations, the effect of dispersal becomes balanced by the effect of selection, and the frequency of the transgene is then about 10% at a distance of 200 m from the boundary. With selective advantage, after some initial generations, a steady state wave front is formed which moves with a speed approximately equal to the prediction of Fisher's analysis.

Another result, and perhaps the most important one, is the small difference between the two scenarios in the predicted frequency of the transgene during the first few generations. After five generations the difference between the position of the gene frequency clines for selective advantage and disadvantage (represented by the two leftmost curves in Figs 1a and 1b) is very small and only about 10 m. Only much later does the effect of selection become apparent. The main reason for this is that when $p_{t,i}$ is kept at 1 in the transgenic part of the population, dispersal will effectively occur in only one direction across the boundary into the the natural part of the population. This form of one-way migration will dominate the dynamics of the gene frequency cline during the first generations, regardless of whether the selection coefficient is positive or negative. Only several generations later is either an equilibrium attained between dispersal and selection, or a steady state wave front moving "with it's own help" is established.

Discussion

According to theory and existing estimates of some of the important parameters, the spread of transgenes into populations of wild relatives is likely to occur rather slowly. This has to be kept in mind when designing monitoring programmes and when evaluating the information that such monitoring programmes produce. Taken together with the fact that genetically engineered plants have been in use for at most two decades, that is, just a few generations, the absence of studies documenting the spread into populations of wild relatives and related environmental effects is perhaps not very surprising. It is well known, however, that at least in panmictic populations, even though the rate of increase in gene frequency may be slow, the probability of fixation of a gene rapidly tends to one for only slightly selectivly favoured genes [8, p. 425].

Even though the spread through large geographic regions may be slow, we should still be concerned with possible long-term effects, if the use of genetically modified organisms is to be sustainable, and the most critical parameter determining the long-term spread of a transgene is its selection coefficient s. The result that the gene frequency cline during the first generations may depend only weakly on the coefficient of selection, suggests that a monitoring programme, sampling near a permanent transgenic plant field, will produce little information about the coefficent of selection. The models discussed in this paper are also fully deterministic, whereas a gene frequency cline, in a real population, will be influenced also by local genetic drift which is likely to mask any pattern that may be present in the data obtained.

Empirical evidence [9] and theoretical models for some modes of dispersal [3], suggest that the distributional form of seed and pollen displacements is quite leptokurtic, that is, a large part of seeds and pollen grains are deposited at either relatively short or long distances. Since dispersal of, at least, airborne pollen is essentially a stochastic non-deterministic process, this means that dispersal over any distance always is possible, and this has led some authors to the conclusion that gene flow by, for example, air borne pollen, is very effective also over long distances. Long distance dispersal will typically occur with a low probability, however, and because of competition with large amounts of local pollen, long distance dispersal from some foreign population will in general not be as important as it may seem, because the contribution to the next generation made up by the foreign pollen in such cases will be very slight only. The actual shape of the dispersal distribution can be important, however, in estimation of the dispersal variance σ^2, because the distribution in most cases is observed over a limited study area only.

Acknowledgements
This work is in part based on financial support from the Norwegian Institute for Nature Research and the Directorate for Nature Management.

References

1. Fisher R A (1937) The wave of advance of advantegeous genes. *Ann Eugenics* 7: 355–369
2. Kolmogorov A, Petrovskii I, Piskunov N (1937) *In*: F Oliveira-Pinto, B Connolly (eds): *Applicable mathematics of non-physical phenomena*. John Wiley and Sons, New York, 171–184
3. Tufto J, Engen S, Hindar K (1997) Stochastic dispersal processes in plant populations. *Theor Pop Biol* 52: 16–26
4. Kareiva P, Morris W, Jacobi C M (1994) Studying and managing the risk of cross-fertilization between transgenic crops and wild relatives. *Mol Ecol* 3: 15–21
5. Nurminiemi M, Tufto J, Nilsson N-O, Rognli O-A (1998) Spatial models of pollen dispersal in the forage grass meadow fescue. *Evol Ecol* 12: 487–502
6. Tufto J, S Engen K Hindar (1997) Inferring patterns of migration from gene frequencies under equilibrium conditions. *Genetics* 144: 1911–1921
7. Wright J W (1953) Pollen dispersion studies: Some practical applications. *J Forest* 51: 114–118
8. Crow J, Kimura M (1970) *An introduction to population genetics theory*. Harper and Row, New York
9. Levin D A, Kerster H W (1974) Gene Flow in Seed Plants. *Evol Biol* 7: 139–220

Discussion session 2: Ecological effects of transgenes

Glynis Giddings
There are in fact various reasons why transgenes might persist in a population even if they have a selective disadvantage. In the work of Jarle Tufto, repeated introgression stochastically and heterogeneity could link to the life cycle, thus, transgenes could persist long enough for evolution to elevate GMO's to a different fitness.

Gösta Kjellsson
I have a question to Christian Damgaard. Do you think that is a good idea, if you want to prevent coevolutive processes, to count with migration? I think that migration might bring insensitive types into the population or maybe resistant types. How would you quantify that idea?

Christian Damgaard
I put migration in the model, because there is seed-migration and there is migration of spores and pollen from one population to the other. If that is not the case in the actually scenery that you are modelling, then you are free to put it to zero. If you have reasons to think that there is no migration, migration in the cases I have looked at makes the effect. The effect I was talking about was that an avirulence gene was allowed to hide somewhere in the metapopulation. Migration diminishes that effect. Throwing away migration would increase the differentiation between the different subpopulations, increasing the effect I was talking about. So, dependent upon the ecological situation, you could include migration or exclude migration.

I would keep an open mind on migration anyway.

Gösta Kjellsson
I have one question to Jarle. We discussed this a little bit yesterday. Now, you are looking at the wave front of pollen and pollen dispersal. What about dispersal of seeds and the cones both more stochastically and dispersed by animals or birds. How do you include that when you see it in the practical terms of modelling?

Jarle Tufto
Seed dispersal in nature was included in the model, not the one caused by man.

Gösta Kjellsson
That may show some of the problems with modelling, because they have a limited scope. We have to validate models with real world data, otherwise they remain just helpful concepts of the mind.

Christian Damgaard
Models are no trivial mind-game, because before you get the data it is useful to first formulate what you actually predict would happen, which is a big help in making these initial hypotheses.

Klaus Ammann
Well, for me the modelling is a great advantage by having the character of disciplining the ecologist. Still, we need a balance of observation, modelling and data-collection.

Glynis Giddings
I agree that the "animal" which is most likely to move seeds about the countryside is in fact man. This is a very difficult problem and models have always been criticised for not including enough parameters. But we have the same problem with experiments. (Hassell MP (1975) Density dependence in single species populations. *J Anim Ecol* 42: 693–726)

Jan Carel Zadok
Different ways of dispersal be incorporated into models: Various dual dispersal models have been published in phyto-pathology which are perfectly useful, and if you want triple dispersal you can do it. A word to Glynis

Giddings: I fully agree with the model output survivorship, fecundity and dispersal. But this is only because we have put them into it. This is a word of caution and hopefully useful for further discussions.

Glynis Giddings
Yes, this is a good point. We do not often put things like "the action of man" into these models, we do not put things in like practices in agriculture. We need to look at these parameters, too, although it makes the models quite complicated.

Jim White
I think we should consider seriously the gene-for-gene hypothesis. Traditional resistance genes often fail after some years. What about transgenes? It is likely that pest will overcome transgene resistance too. One scenario that has been discussed is insects overcoming BT genes. Another case is the one with the papaya ring spot (PRSV) coat protein resistant papayas that have been commercialized in the US; there are exotic strains of PRSV to the US that these transgenic plants are susceptible too. This suggests that resistance breaking strains might occur in papayas.

We should also consider Alison Snow's recent statement about escaping herbicide resistance transgenes in colza: In general, however, there are few examples of weeds benefiting from specific fitness-related crop genes. This could be due to several factors – the lack of attention to the phenomenon, the absence of crop genes that confer strong fitness advantage to relatives, or simply the fact that the impact of beneficial genes is not dramatic. (*Snow, A.A. and Palma, P. M. 1997. Commercialization of transgenic plants: potential ecological risks. BioScience 47:86–96.*)

Klaus Ammann
From the point of view of ecology I think it is much better to have a resistance system which is not working so perfectly well. It is rewarding to hear from American entomologists that there is presumably also natural resistance against Bt.

François Pythoud
How can modelling help the work of the regulators when they have to make decisions on an application?

Glynis Giddings
I think there are various positive things we can do. In designing experiments we can address some idea of probabilities. We can learn how many environments we need, how many years of data we need in order to be at least reasonably predictive.

Wolfram Hemmer
I want to add a comment to Mr. Pythoud's question. Something which possibly could help even more for regulatory decisions would be calculation of damage-potential and not only for comparison to calculation of probabilities.

Tom Nickson
Just from my perspective, one comment. I worked with a little bit of modelling as chemist, years ago, and I have seen it now used as an interesting complement to the experiments in biology. But I would stress the point that for a regulator to use a model without any experimental validation, would be a very dangerous thing. I think that the modellers do a very nice job of complementing and rationalising why something is happening, but there is no certainty, you cannot make any decisions without data based on valid experimental methods, and I think that we cannot lose that here.

I would really caution any kind of a statement where the model doesn't fit the data.

Klaus Ammann
I must say that data can also be wrong. If they have been collected in the wrong way, besides miscalculation or things like that or worse, if inappropriate data have been collected. And I think also there modelling can help. I agree fully: it should discipline the regulatory processes in a way that risk assessment should be channelled through the interaction with good modellers.

Christian Damgaard
I do not think there is controversy between experiment and model. Some of these systems are so complex, with the genetical rearrangements and the coevolution consequences, that our intuition does not hold very well. Why not let our intuitions be trained and helped by the tool of modelling?

Klaus Ammann
I think that was a good final sentence. And thanks for the high level of discussion.

Session 3: Short-term, long-term effects and standardisation of limits

Methods for Risk Assessment of Transgenic Plants
III. Ecological risks and propspects of transgenic plants
K. Ammann, Y. Jacot, V. Simonsen and G. Kjellsson (eds)

Short-term effects, long-term effects and standardisation of limits

Philip J. Dale

John Innes Centre, Colney Lane, Norwich NR4 7UH, UK

Introduction

The risk assessment process is supported by data from many sources.A criticism of risk assessment data from short-term and small scale experiments is that some of the information they provide cannot be extrapolated to assess the impact of the long-term and large scale production of transgenic crops. Currently, a range of transgenic crop plants is being approved for large scale agricultural production. A rare hybridisation event, for example, between a crop plant and a related species is unlikely to be detected in a 1-hectare experiment carried out for 1 year, but could be significant when the crop is grown on 100 000 hectares over a 10-year period. The aim of this paper is to consider how we cope with assessing the impact of crops to be grown long-term and widespread within agriculture. How do we bridge the gap between data on short-term effects of transgenic plants, to predict their possible impacts when grown extensively over many years, and is it possible to set standards or criteria to help with this process?

Short-term experiments and long-term releases

When a transgene is introduced for the first time, it is usual to test its function and stability under contained glasshouse conditions. The evaluation process then usually continues in small scale field experiments (50–500 m^2). Depending on the nature of the crop, the transgene and the environment, there may be a requirement to use “field containment” conditions. These can include genetic isolation measures such as physical separation from sexually compatible plants, the use of barrier crops, the removal of sexually compatible wild species, and other regulatory restrictions may be imposed. This initial testing is usually on one site and in one environment. As the evaluation process continues, the plot size and number are increased and will eventually include several different experimental sites and environments. Evaluation experiments of these kinds are important for learning about the expression and stability of the transgenes in particular lines (specific transgene insertion events) and to give an indication of gene expression levels and tissue specificity in field environments; however, they are unlikely to give comprehensive information on all of the possible impacts of those transgenic plants when grown extensively in agriculture.

Various potential impacts are likely to be affected by the length and scale of transgenic plant cultivation. These impacts include: gene transfer to other plants by hybridisation; non-target effects on friendly organisms within the environment; genetic interactions between different transgene constructs; interactions between transgenes and resident genes in different environments; changes in virulence of pests and pathogens in response to the use of resistance genes; invasiveness of transgenic plants and their progeny in natural habitats; and persistence of transgenic plants and their progeny in agricultural habitats.

How do we bridge the gap?

It is sometimes tempting to make generic responses to risk assessment. In general, these kinds of reactions are not helpful to the risk assessment process.

Some examples of generic responses are as follows:

1. All gene transfers across sexual barriers are unnatural, undesirable, unacceptable and a kind of genetic pollution. This overlooks the fact that methods to overcome sexual barriers have been used in conventional breeding for many decades, by the use of *in-vitro* ovary and embryo culture techniques.
2. Disapproval of the principle of introducing transgenes to make plants tolerant to particular herbicides. Again, conventional plant breeding has been used to produce herbicide tolerant crops for many decades.
3. All non-target effects on friendly organisms within the environment are undesirable and unacceptable. But, using insect control as an example, whatever method is employed to limit aphid damage, whether sprays, transgenes or removal of aphids by hand, all are likely to have an impact on ladybird predators.

In considering how we bridge the gap between short- and long-term impacts of transgenic plants, gene transfer to related plant species by sexual hybridisation will be used to illustrate some principles. The likelihood of transgene dissemination to other plant species can be divided into qualitative and quantitative effects. Plant species can usually be classified into those that are sexually compatible, and those that are not sexually compatible with crop species (qualitative effect; see for example [1, 2]). Most species in nature are sexually incompatible with crop plants, and therefore for those species the possibility of gene transfer can be disregarded during risk assessment.

When gene transfer between crop species and related plant species is known to occur, there can be considerable quantitative variation in the likelihood of that hybridisation under field conditions [1,3–5]. Even when a crop is considered to be largely sexually incompatible with a related species, it is important to consider the extent to which genetic and environmental variation might influence the state of sexual incompatibility. It is widely known from attempts to introduce genes into crops from related species or genera in conventional plant breeding that the ease of sexual transfer of a gene into a crop is usually influenced by the genotypes of the donor and recipient species, and the environments the plants are grown in. When hybrids are formed, there are also genetic and environmental influences on the success of those hybrids.

An assessment of the likelihood of transgene transfer to related species is an important part of the risk assessment process before permitting an experimental release of transgenic plants and, more importantly, before allowing commercialisation of a transgenic plant variety. If hybridisation is known to be possible under field conditions, it is usual to assume that hybrids will form during agricultural practice. The emphasis of the risk assessment is then transferred to considering the consequences of that hybridisation. For example, if herbicide tolerance is transferred to weed species, what is the likely impact of that transfer on the agronomic environment, and are there agronomic and agricultural practices that can cope with particular weed species that have become tolerant to specific herbicides. Similarly, if pest or disease resistance is transferred to plants in natural habitats, what is the likely impact of that transfer, and will the transgenes confer a significant advantage in those habitats [6]?

Can we define criteria or standards?

It is appealing to attempt to define standards of acceptable and unacceptable impacts of transgenic plants on the environment. This might make it easier to monitor and police the widespread use of transgenic plants in agriculture. However, before it is possible to do this, it would be necessary to define, in rather precise terms, what is meant by impact. The responsibility provided within the European Union by the Directive (90/220) governing the release and marketing of genetically modified organisms, requires a consideration of impact on "human health and the environment". It is also important to consider the extent to which this includes impact on the closely managed "agricultural environment". To what extent should the regulatory process be involved with orchestrating agronomic practices within agriculture? In most cases it will be very difficult to define precise, sensible and policeable criteria of the kind that may be favoured by some regulatory authorities.

Developing criteria

It will be very difficult and often impractical to define generic kinds and levels of unacceptable impacts on the environment. For example, it may be tempting to say that to kill more than 20% of earthworms or ladybirds over a given period and habitat, is unacceptable and represents a significant adverse effect on a friendly organism. It is, in my view, important that decisions on criteria for acceptable and unacceptable impacts must be specific to particular crops, genes and environments. For example, the consideration of a particular herbicide tolerance gene in oilseed rape must take into account the likelihood and consequences of the growth of herbicide tolerant crop plant volunteers, the transfer of that gene into specific weed species, the likelihood of multiple resistance developing and the agronomic practices that are employed or might be employed for that crop. It is necessary to have specific criteria for particular crops and situations, and solutions to deal with them.

A particularly challenging issue relevant to the question of developing criteria, is the extent to which any negative impact should be balanced by consideration of benefit. For example, if an insect resistant crop that reduces numbers of ladybirds by 20% in a given period and habitat, also reduces the need to use a persistent insecticidal spray, does this influence the final decision on acceptability for commercialisation? Within the EU regulatory process the overwhelming emphasis in risk assessment is to minimising adverse effects on the environment, with negligible (although there is some variation among the 15 EU countries) consideration of potential benefits.

Although the development and adoption of standardised criteria for acceptability and unacceptability will generally be very difficult in practice, it is likely to be feasible for certain specific cases. For example, in conventional plant breeding, certain standards of seed purity are required for the production of breeders' basic seed during seed multiplication (<0.1% contamination; [5]). Similarly, standards of seed purity are required for the production of certain oil types of varieties of oilseed rape. It is not difficult to envisage precise standards of contamination being set for the production of crops grown for industrial processing or pharmaceutical extraction. It is also reasonable to set standards of contamination for the production of food crops grown in the vicinity of transgenic crops for industrial and pharmaceutical processing, where a risk assessment shows this to be necessary or desirable.

Strict standards and criteria for impact on the environment will, therefore, mostly be very difficult to define and impractical to operate. If the transfer of a specific gene to a natural population is genuinely considered to present a significant hazard, then genetic isolating mechanisms should be considered or the transgenic plant release should not be allowed.

How do we decide?

How do we carry out the best possible assessment of environmental impact for the widespread use of a particular transgenic crop in agriculture? It is important to emphasise that risk assessment is an imperfect science. Assessments and measurements of impact on the environment could be carried out for the next 50 years, and still not provide evidence to satisfy all concerns about the long-term impact of genetically modified crops. The fundamental initial choice is, either we conclude that the potential impact of genetically modified crops is unacceptable and therefore ban or inhibit their development; or we accept that genetic modification has valuable attributes and is capable of making a useful contribution to agriculture. In my view the arguments against inhibiting the development of genetically modified crops for agriculture are:

1. It would make a generic negative judgement about a whole area of scientific enquiry and development.
2. It would prevent the development of crops that have the potential to make a very significant contribution to human health and the environment throughout the world.
3. Transgenic crops are already in widespread agricultural production in certain parts of the world (principally North America), and because of regulatory harmonisation and trade agreements, it would be very difficult to inhibit significantly the importation of genetically modified crops and their products into Europe.

Conversely, if we accept the principle that genetically modified crops have the potential to make a very valuable contribution to European and world agriculture, then we must accept that risk assessment and scientific analysis of impacts over the long-term will inevitably always be imperfect and incomplete. All aspects of agriculture and human existence carry a degree of risk. It is also important that we should not penalise transgenic plant breeding simply because of its novelty.

The regulatory process that assesses the acceptability of the commercial production of genetically modified crops must draw on all the experience and evidence that is available, including the following:

1. Evidence from the experimental releases and breeding trials carried out on modified crops by plant breeders, and those involved in variety evaluation and comparison. In the later phases of plant breeding and evaluation, the process includes large scale, multi-site and multiple environment evaluation plots. If the regulatory authorities require additional environmental impact data for commercial approval, then this field evaluation phase can be an opportune time to gather such information.

2. It is important that we draw on the experience of conventional plant breeding over many years. Many of the transgenic varieties currently being produced have phenotypes very similar to those that have been modified for many decades in conventional plant breeding. If the genetically modified plants are tolerant to particular herbicides, then it is appropriate to compare and contrast them with other herbicide tolerant varieties of the same crop. This principle also applies to pest and disease resistance and a wide range of other characters.
3. The concept of substantial equivalence is widely used in the assessment of novel foods, including those from genetically modified organisms. Although there is debate about the criteria that are used to define substantial equivalence – the principle of comparing the environmental and other impacts of genetically modified plants with plant varieties well established in agriculture, is a compelling one.
4. There is extensive experience of ecological experimentation and observation relevant to assessing the characteristics that influence plant persistence and invasiveness, in natural and managed habitats.
5. An adjunct to all of these is monitoring. It is important to take the challenge of monitoring very seriously and to consider how this can be done in practice, who might do it and whether it is appropriate for some directed monitoring to be carried out by farmers and the agricultural extension services. Within the United Kingdom the National Farmers Union (NFU), the British Society for Plant Breeding (BSPB) and the United Kingdom Seed Trade Association (UKASTA) have combined to develop codes of practice for use in farming and are currently debating codes of practice appropriate for the cultivation of specific genetically modified crops.

Summary and conclusions

Attempting to extrapolate evidence from short-term release experiments with genetically modified crops, to estimate the long-term impact of widespread agricultural production, presents a significant challenge. Generic reactions to impacts or perceived impacts on the environment, such as the concept of genetic pollution, or blanket judgements about non-target effects on friendly organisms within the environment, are usually not helpful in decision making. Risk assessment needs to be specific to the crop, the transgene and the environment. It is attractive to try to develop criteria and standards of environmental impact that are policeable and can be monitored. In practice it is difficult to define in precise terms what are acceptable and unacceptable impacts. There are differing risk assessment considerations depending on whether the potential impacts are on the natural environment or the agricultural environment. There is currently debate about the extent to which risk assessment and regulatory oversight should include management of the agricultural environment which is already a highly managed habitat. There is also discussion about whether environmental impact should include an assessment of benefit. An analysis of benefit is appealing, but we should also ask: benefit to whom? Criteria should and will be definable in certain specific cases, but in general, describing standard criteria for environmental impact will be difficult and impractical. Risk assessment and environmental impact assessment will of necessity provide incomplete answers. All aspects of human existence carry some level of risk. Decisions on the long-term use of genetically modified crops can only be made by drawing on the extensive experience of breeding, agricultural practice and ecological observation. There are compelling arguments for the long-term monitoring of genetically modified crops used in agriculture, but it is important to consider who should be responsible for monitoring, how should it be done and will it be informative in practice?

Acknowledgements
I thank the BBSRC, DETR and MAFF for their support.

References

1. McPartlan HC, Dale PJ (1994) The transfer of introduced genes from field grown transgenic potatoes to non-transgenic potatoes and related solanaceous species. *Transgenic Res* 3: 216–225
2. Scheffler JA, Dale PJ (1994) Opportunities for gene transfer from transgenic oilseed rape (*Brassica napus*) to related species. *Transgenic Res* 3: 263–278
3. Conner AJ, Dale PJ (1996) Reconsideration of pollen dispersal data from field trials of transgenic potatoes. *Theor Appl Genet* 92: 505–508
4. Scheffler JA, Parkinson R, Dale PJ (1993) Frequency and distance of pollen dispersal from transgenic oilseed rape (*Brassica napus*). *Transgenic Res* 2: 356–364
5. Scheffler JA, Parkinson R, Dale PJ (1995) Evaluating the effectiveness of isolation distances for field plots of oilseed rape (*Brassica napus*) using a herbicide-resistance transgene as a selectable marker. *Plant Breeding* 114: 317–321
6. Raybould AF, Gray AJ (1993) Genetically modified crops and hybridisation with wild relatives: a UK perspective. *J Appl Ecol* 30: 199–219

Elimination of agrobacteria from transgenic plants

Jörg Landsmann, Elke Graser and Anja Matzk

Biologische Bundesanstalt für Land- und Forstwirtschaft, Institut für Biochemie und Pflanzenvirologie, D-38104 Braunschweig

Summary. Genetically engineered agrobacteria are routinely used to transform crop plants. These agrobacteria can obviously persist within the transgenic plants although they are not tumorgenic [1–4]. Our attempts to eliminate the agrobacteria concentrate on the regeneration of secondary shoots from infected tobacco plants. Application of the antibiotics cefotaxim or carbenicillin during the regeneration process did not result in sufficient reduction of the bacterial contamination. Shoot tip culture of apical meristems, however, resulted in a large percentage of agrobacteria-free regenerated plants (within the detection limits). The degree of plant contamination with agrobacteria was determined by Polymerase Chain Reaction (PCR).

Implications for the assessment of human or environmental risks associated with the release into the environment of genetically engineered plants containing persisting genetically engineered agrobacteria, however, seem relevant only in specific cases, e. g. if host ranges of pathogens or selective advantages of microorganisms by means of resistance traits might be altered or increased.

Leaf disk transformation of tobacco

Leaves from sterile grown *Nicotiana tabacum* var. W38 were cut into strips and transformed with *Agrobacterium tumefaciens* LBA4404 [2, 5] containing derivatives of the binary vector pLX222 [6]. Selection with kanamycin (100 mg/ml) and agrobacterial counterselection with Claforan (Hoechst) (250 mg/ml) were maintained for 4 weeks. No agrobacteria grew on this medium.

During shoot tip propagation [7], i.e. more than 5 weeks after release of Claforan counterselection and more than 9 weeks after transformation, occasionally bacteria could be detected growing out of the cut stem site of the transgenic shoots.

Identification of the reisolated bacteria

All bacteria isolated 6 months after plant transformation from the transformants were positive in a ketolactose test [8] indicating they were *A. tumefaciens*. The bacterial isolates were tested in a BIOLOG test (Biolog Inc., Hayward, CA, USA) and could be clearly identified as *Agrobacterium* spp.

Two of 18 isolates tested had lost one of their antibiotic resistance markers.

Plasmid preparations from all 60 isolates tested resulted in plasmid DNA of the expected size, although the vectors were generally not completely stable in *A. tumefaciens* LBA4404. Restriction analyses showed the restriction pattern of the original binary vectors. Southern hybridisation with a vector probe detected the correct fragments although additional rearrangements had taken place.

Functional integrity of the reisolated agrobacteria

In order to check the integrity of the constructs two of the reisolated *Agrobacterium* strains were used to transform tobacco with the leaf disk method. Six regenerated plants of each transformation experiment were grown to the rooting stage. The level and variability of GUS activity of the regenerated plants proved to be comparable to the primary set of transformants [2, 9].

Localisation of agrobacteria in the plant tissue

Two methods were applied to localise the persisting agrobacteria in the plant tissue: tissue print immunoblotting (TPIB) and scanning electron microscopy (SEM) [10].

Examination of the leaf and stem samples of *in vitro* transgenic tobacco plants by SEM revealed bacteria in all five plants tested. The external surfaces were always found to be colonised by bacteria but their distribution was not uniform. The bacteria mainly appeared in colonies, often associated with an extracellular matrix. Within tissues exposed by freeze-fracture, bacteria were rarely observed. Bacteria could be found in the epidermal layer, the intercellular spaces beneath the epidermis and within the vascular tissue. No bacteria were detected on the internal surfaces of the roots.

TPIB was performed according to [11]. The *in vitro* cultivated tobacco leaf and stem tissues were cut and the freshly exposed surfaces pressed onto nitrocellulose membrane. Bacterial antigens were detected by means of alkaline phosphatase-labeled *A. tumefaciens* C58C1-specific antibodies. Agrobacteria could always be localised in stem sections and frequently in leaf sections. The external surfaces of tissues often showed higher density of agrobacteria.

When testing tissue from transgenic tobacco plants which had been grown in soil under non-sterile conditions, there was a high concentration of agrobacteria in the stem base of the plants, but no signals in the middle and upper part of the stem. Root material was not tested with TPIB.

Cefotaxim (Claforan) or carbenicillin are not sufficient for elimination of agrobacteria

Pieces of leaves from transgenic *in vitro* tobacco plants, still harbouring recombinant agrobacteria, were incubated on MS-medium to regenerate shoots. During the regeneration process Claforan 250 µg/ml or carbenicillin 500 µg/ml were applied. Fifty shoots of each experiment were analysed. One of the shoots developed agrobacteria when transferred to antibiotic-free medium. The majority of the shoots were positive in PCR analyses with primers for the agrobacterial chromosomal ros gene, i. e. they showed amplification products of the expected size on ethidium bromide stained agarose gels [12, 13].

Shoot meristem culture reduces agrobacterial contaminations

Shoot tips were prepared from transgenic *in vitro* tobacco plants and used to regenerate calli and shoots [14]. 50% of the regenerated calli from 57 meristem preparations >0.5 mm in diameter and 85% of calli regenerated from 139 meristems <0.5 mm in diameter did not develop agrobacterial colonies on the medium (Tab. 1). DNA preparations of these calli were submit-

Table 1. Regeneration of calli from shoot apical meristems of transgenic tobacco plants and detection of agrobacterial colonies on the surface of the cultivation medium

Meristem size ∅	∅ ≤ 5 mm	∅ > 5 mm	total
Number of total meristem preparations	149	59	208
Number of successful callus regenerations	139	57	196
Preparations with agrobacterial colonies	22	29	51
Preparations without agrobacterial colonies	127	28	155

Table 2. ros-gene PCR with DNA preparations from calli regenerated from shoot apical meristems of transgenic tobacco plants

Successful detection of agrobacterial colonies (see Tab. 1)	Yes	No
Number of calli tested in ros-PCR	27	117
Number of calli positive for agrobacteria according to ethidium bromide stained agarose gel of ros-PCR	24	0
Number of calli positive for agrobacteria according to hybridisation of PCR-blot with ros gene	27	34

ted to PCR analyses including hybridisation with the ros gene of the *Agrobacterium* chromosome as hybridisation probe [3, 15]. Seventy-one percent of the calli showed negative results in the ros hybridisation of the PCR gels (Tab. 2). Primers for the tet gene of the binary vector pLX222 could not detect visible amplification products in PCR analyses with the bacteria-free candidates. The tet gene is located outside the T-DNA region and should not be transferred to the plant chromosome upon plant transformation.

Tobacco seeds exclude agrobacteria

Several hundred seeds from different individual tobacco transformants were aseptically germinated on MS-medium. The emerging plantlets were cut into pieces and incubated at 28 °C on LB-medium agar. After 10 days still no agrobacteria were visible. The remaining plantlets were grown for a further 4 months, with no bacteria appearing on the MS-medium.

Seedlings from different transgenic tobacco lines were submitted to PCR analyses with the ros and the tet primers. 100% of 46 tet-PCR analyses of two transgenic lines (comprising 2–3 pooled seedlings each) and all 20 tet-PCR analyses of leaves from individual plantlets of another transgenic line were negative [12].

Discussion

Engineered agrobacteria residing within transgenic plants in the field are capable of gene transfer to other bacteria in the environment. An uncontrolled spread of genes from non-bacterial kingdoms to the microflora of natural and agricultural ecosystems is not desired to date. Thus plants released into the environment should be free from engineered agrobacteria [16–19].

It should be noted, however, that in the vast majority of cases no significant impact and thus no risk for the environment would be connected with such a horizontal gene transfer.

Implications for the assessment of human or environmental risks associated with the release into the environment of genetically engineered plants containing persisting genetically engineered agrobacteria, however, seem relevant only in specific cases; e. g infective plant virus genomes on the agrobacterial T-DNA could overcome natural infection barriers for that virus and thus theoretically increase the host range of that pathogen; e. g natural transfer of antibiotic resistance genes from persisting agrobacteria to pathogenic microorganisms could increase the selective advantage of those microorganisms and thus under specific conditions impede medical therapies.

Agrobacteria persisting in plants after natural infections [20–22] as well as after plant transformation through genetic engineering appear to be defying even severe attempts of elimination. Nevertheless, seed transmission of agrobacteria can probably be denied at least for tobacco. Notwithstanding, tests proving the bacteria-free state of a plant depend upon the detection limit and need to be standardised for general use [23–25]. The detection limit for visible PCR products on ethidium bromide stained gels lay at <100 bacteria per reaction tube. Additional hybridisation could detect even one bacterium. However, when calculating the necessary dilutions (PCR does not work in concentrated extracts) and extrapolating to the plant material assayed it came down to a minimum of 750 agrobacteria which escape detection.

Acknowledgements
We would like to thank Gundi Püster for excellent technical assistance.

References

1. van der Hoeven C, Dietz A, Landsmann J (1991) Agrobacteria shown to reside in transgenic plants. *Nachrichtenbl Deut Pflanzenschutzd* 43: 249–251
2. van der Hoeven C (1992) *Untersuchungen zur Variabilität von gentechnisch veränderten Tabakpflanzen mit Resistenz- und Reportergenen.* Ph.D. Dissertation. Universität Hannover, Germany
3. Zweigerdt R (1993) *Nachweis persistierender Agrobakterien in transgenem Kartoffelmaterial.* Diplomarbeit. Universität Braunschweig, Germany
4. Mogilner N, Zutra D, Gafny R, Barjoseph M (1993) The persistence of engineered *Agrobacterium tumefaciens* in agroinfected plants. *Mol Plant Microbe Interact* 6: 673–675
5. Horsch RB, Fry JE, Hoffmann NL, Eichholtz D, Rogers SG, Fraley RT (1985) A simple and general method for transferring genes into plants. *Science* 227: 1229–1231
6. Landsmann J, Llewellyn D, Dennis ES, Peacock WJ (1988) Organ regulated expression of the *Parasponia andersonii* haemoglobin gene in transgenic tobacco plants. *Mol Gen Genet* 214: 68–73
7. Murashige T, Skoog F (1962) A revised medium for rapid growth and bio assays with tobacco tissue cultures. *Physiol Plant* 15: 473–497
8. Bernaerts MJ, De Ley J (1963) A biochemical test for crown gall bacteria. *Nature* 197: 406–407
9. Gould JH, Smith RH (1989) A non-destructive assay for GUS in the media of plant tissue cultures. *Plant Mol Biol* 7: 209–216
10. Matzk A, Mantell S, Schiemann J (1996) Localization of persisting agrobacteria in transgenic plants. *MPMI* 9/5: 373–381
11. Kaufmann A, Koenig R, Lesemann D-E (1992) Tissue print immunoblotting reveals an uneven distribution of beet necrotic yellow vein and beet soil-borne viruses in sugarbeets. *Arch Virol* 126: 329–335
12. Landsmann J, Graser E, Riedel-Preuß, A, van der Hoeven C (1995) Experiments to eliminate agrobacteria persisting in plants. *Nachrichtenbl Deut Pflanzenschutzd* 47: 240–244
13. Landsmann J, Graser E, Riedel-Preuß, A, van der Hoeven C (1996) Can agrobacteria be eliminated from transgenic plants? *In*: ER Schmidt, Th Hankeln (eds): *Transgenic organisms and biosafety,* 71–76

14. Graser E (1994) *Eliminierung von Agrobakterien aus gentechnisch veränderten Pflanzen*. Diplomarbeit. Universität Göttingen, Germany
15. Dsouzaault MR, Cooley MB, Kado CI (1993) Analysis of the ros repressor of *Agrobacterium* virC and virD operons – molecular intercommunication between plasmid and chromosomal genes. *J Bacteriol* 175: 3486–3490
16. Casper R, Landsmann J (1992) Summary of Results. *In*: RCasper, J Landsmann (eds): *Proc. Sec. Int. Symp. on The Biosafety Results of Field Tests of Genetically Modified Plants and Microorganisms. Goslar.* BBA, Braunschweig, 13–19
17. Landsmann J (1992) Organisms altered by gene technology and released into the environment. An examination of the risks. *Plant Res Dev* 35: 73–79
18. Landsmann J (1995) Release of Transgenic Plants. Regulations and Developments. *In*: I Potrykus, G Spangenberg (eds): *Gene Transfer to Plants*. Springer Verlag, Heidelberg 32: 340–349
19. Barrett C, Cobb E, McNicol R, Lyon G (1997) A risk assessment study of plant genetic transformation using *Agrobacterium* and implications for analysis of transgenic plants. *Plant Cell Tissue Organ Cult* 47: 135–144
20. Lehoczkky J (1971) Further evidence concerning the systematic spreading of *Agrobacterium tumefaciens* in the vascular system of the grapevines. *Vitis* 10: 215–221
21. Stellmach G (1990) Experiences with the isolation of latent tumorgenic agrobacteria from grapevines. *Nachrichtenbl Deut Pflanzenschutzd* 42: 151–153
22. Bauer C, Schulz T F, Lorenz D, Eichhorn K W, Plapp R (1994) Population dynamics of *Agrobacterium vitis* in two grapevine varieties during the vegetation period. *Vitis* 33: 25–29
23. Cooley MB, D'Souza MR, Kado CI (1991) The virC and virD operons of the *Agrobacterium* Ti plasmid are regulated by the ros chromosomal gene: Analysis of the cloned ros gene. *J Bacteriol* 173: 2608–2616
24. Dong I-C, Sun C-W, Thies KL, Luthe DS, Graves JCH (1992) Use of polymerase chain reaction to detect pathogenic strains of *Agrobacterium. Phytopathology* 82: 434–439
25. Ponsonnet C, Nesme X (1994) Identification of *Agrobacterium* strains by PCR-RFLP analysis of pTi and chromosomal regions. *Arch Microbiol* 161: 300–309

Assessment of long-term environmental impacts of transgenic trees: Norway spruce as a case study

Bjørn Å. Tømmerås and Kjetil Hindar

Norwegian Institute for Nature Research (NINA), Tungasletta 2, N-7005 Trondheim, Norway

Introduction

The aim of this paper is to contribute to the development of strategies for assessing ecological impacts of transgenic tree species. Norway spruce (*Picea abies* (L.) Karst.) plays a key role in boreal ecosystems from Norway to Siberia and is also an economically very important species. A hypothetical release of transgenic spruce is selected for this case-study exercise, modified from [1].

Risk assessment and some/their limitations

Norway spruce plays a key role in natural ecosystems by, for example, changing the soil quality (chemically and functionally), the ability to create structure and the importance for the forest dynamics in the boreal zone. Interbreeding between cultured and natural populations is common in the species. Thus, Norway spruce is a very relevant model system for evaluating environmental impacts of genetic modification of a key species which features both cultured and natural populations, i.e. a semi-domesticated species.

There is no established definition of "long-term" in connection with assessing environmental impacts. We have followed a pragmatic definition of 10 to 100 generations as a guideline [3]. The generation time for Norway spruce is approximately 25 years, and the time scale for long-term environmental impacts of release of transgenic trees can therefore be hundreds of years. The Ecological Society of America has proposed the following principle for risk evaluations when genetically modified organisms are released to the environment [4]: "Genetically modified organisms should be evaluated and regulated according to their biological properties (phenotypes), rather than according to the genetic techniques used to produce them." Others have pointed out that the technique used in the genetic engineering must also be subject to assessment, for example with respect to regulating gene expression and uncertain time lag effects.

There will always be an element of biological "gambling" involved in the release of genetically modified and other (non-modified) organisms. Our genetic and ecological knowledge is inadequate to allow us to give *a priori*, precise risk assessments or reliable predictions of aimed success for releases [5–7]. Lack of knowledge, especially on topics related to the biodiversity and function of forest ecosystems, is partly responsible for uncertainties in predicting impacts by releasing transgenic spruce. Moreover, some short-term and especially long-term effects are more or less unpredictable due to stochasticity in climatic and biological conditions and the specific conditions in the area at the time of release, as experienced from impacts of some introduced plant species [2].

An important principle for risk assessment is that the environmental effects of genetically modified organisms must be evaluated on a case-by-case basis through a stepwise procedure that includes fully enclosed pilot studies [8]. There is broad international agreement about this principle [9, 4]. But even a risk assessment based on the principles of "case-by-case" and "step-by-step" has limitations such as (1) we have to use short-term experience to assess long-term impacts, (2) we can only test the possible problems we are aware of, (3) an inability to prove any impact can be caused by problems with the method or that the investigation does not cover all the relevant effects, and (4) for organisms with long generation time as forest trees time is a limiting factor. Moreover, it has been shown in many cases of deliberate introductions that the persistence of the released population largely depends upon how many individuals were released [10] and that changes in tree invasiveness after several generations occur [11]. Therefore, the results of small-scale experiments need not translate directly to large-scale releases. Furthermore, in a widespread organism as Norway spruce experience from field trials with transgenics in one region, cannot be applied to risk assessments of release in other regions without explicit tests of the effect of environmental and genetic variation on the outcome of the release. Collectively, these considerations mandate a way of thinking about releases that must incorporate a precautionary approach as a guideline for decisions.

Consideration of environmental effects

The environmental effects may be divided into three broad categories [12]: (1) effects caused by the genetically modified organism itself ("invasion potential"), (2) effects resulting from dispersal of genes from the genetically modified organism to other organisms (intra- and inter-species) in the environment ("gene flow"), and (3) altered practice in the use of an organism because of the genetic modification ("altered use"). Tiedje et al. [4] have provided a detailed list of questions which can be used as a checklist for identifying releases on a subjective scale from low to high risk, a list we have modified.

Invasion potential

Some key questions regarding the invasion potential of a genetically modified organism are: (1) is the expression of the genetic modification well documented and understood? (2) Do the genetically modified trait represent an ecological novelty? (3) To what degree is the host organism domesticated? (4) Is the organism a key species in the environment? (5) Is the release environment "contained" by natural barriers for spread of the organism, or can the spread be controlled by human intervention? And (6) how well can the dynamics of the release environment be simulated prior to, and monitored after, release?

Transgene expression

Increased predictability of the phenotypic expression of the genetic modification in different environments and under different environmental conditions means that the GMO (Genetically Modified Organism) is safer to the environment. Some factors affecting the stability of transgene expression have to be focused. Time-lag effects following changes in gene expressions are unpredictable and several decades are for Norway spruce needed to discover, for example, selective advantage. Both the long generation time and reproductive period highlight the need of stability to reduce uncertainties about this species.

Ecological novelties

Transgenic organisms are claimed to pose greater ecological risks than conventionally cultured organisms, particularly if the transgene has not previously been tested against the genetic and environmental background of the organism (e.g. a flounder gene introduced to spruce) [4].

Domestication

The degree of domestication of an organism often reflects how dependent it is on human support for survival and reproduction. Fully domesticated species are expected to pose less risk to the environment, because any unwanted effects can be controlled by human intervention. The semi-domesticated species Norway spruce grows both in natural stands and in plantations.

Key species

Some species are ecologically more important than others. It is clear that genetically modified varieties of these species will have greater ecological impacts than modified species which play a lesser ecological role. Ecologically important species have been termed "keystone species" which may be defined as: *a species is keystone if its experimental removal (or introduction) causes major ecosystem-level changes in structure, dynamics or nutrient flows* [3]. No species in the boreal zone is as important for the function, structure and dynamics and the whole biodiversity of the boreal forests as Norway spruce.

Natural spread and its control

Releases of genetically modified organisms are considered safer if the organism has a small area within which it can survive and reproduce, and if limited dispersal capability (or an environmental barrier) prevents it from reaching other such areas, should they exist. Norway spruce poses potential risks by existing naturally in large areas, and for which dispersal capabilities (or lack of environmental barriers) can make rapid range expansion possible following, say, genetic modification or environmental change.

Simulation and monitoring

The ecosystem performance of many small, short-lived organisms can be simulated with good ecological realism in the laboratory. For transgenic varieties of these organisms, the possible environmental effects may be well known following evaluation of their performance in a "step-by-step" fashion. The performance of many large, long-lived organisms such as spruce, on the other hand, cannot easily be simulated, because in order to obtain ecological realism, the scale of the experiment would itself demand uncontained conditions.

Gene flow

Some key questions regarding the effects of gene flow from a genetically modified organism are: (1) Does the organism have close relatives (i.e. natural or semi-natural populations of the same or closely related species) in or near the release environment? (2) How large is the natural gene flow among populations of the species? And (3) does the release environment put a selective premium on the introduced gene?

Spruce seeds are mostly produced by outcrossing, but some self-fertilisation occurs. Sexual maturity generally is reached at an age of 15–30 years. Seeds are dispersed mainly by wind and rather few are spread over larger distances. The gene flow by means of pollen dispersal is effective in Norway spruce.

Close relatives

By close relatives, we mean everything from semi-natural or wild populations of the species undergoing genetic modification, to other species of the same genus and even other genera with which the genetically modified organism can hybridise [13, 14]. The potential for introgression of the transgene into wild populations will be highest for situations where the genetically modified organism coexists with conspecific populations that form a natural part of the surrounding environment as for GM-spruce. When released in the boreal zone any release of not 100% sterile genetically modified spruce normally are going to interbreed with natural populations. Natural hybridisation can also occur involving other variants of *P. abies* such as *obovata, jeoenzis* and *koraensis.*

Natural gene flow

The level of natural gene flow in a species can be used as a yardstick for predicting the spread of genes from genetically modified variants of the same species. Several aspects of the natural population structure are important; for example in a highly subdivided population, a transgene can become locally abundant even if there is selection against it, but it will rarely spread far. In contrast, a transgene may spread far into a highly connected population but will rarely reach a high frequency in any location [15]. All these circumstances are present in the more or less heterogeneous landscape of connected spruce forests in the boreal zone. At any rate, it is difficult to precisely predict gene flow from experimental studies of the dispersal of pollen and seeds. For assessing more long-term gene flow, estimates based on studies of genetic differences between extant populations may be appropriate [16].

Selective advantage

Fixation of the transgene in natural populations is almost inevitable if the transgene poses a selective advantage to its carrier. Hence, one crucial piece of information is whether or not the genetically modified trait has a selective advantage in the wild. Norway spruce is, due to its large genetic variability, able to colonise areas with quite different physical and climatic conditions as well as compete with other tree species. Selective advantages can therefore normally not be predicted over its whole distribution area in the boreal zone. Selective advantages in one type of area can be disadvantageous or neutral in another.

Transgenic traits that are considered to be important to assess, especially when released in the area of natural range, include those which increase the environmental tolerance of the species. This tolerance can relate to both the abiotic (e.g. climate) and biotic environment (e.g. competing species), and have effects on the abundance (e.g. carrying capacity) or distribution of the organism. Norway spruce as a key species and the actual GM (Genetically Modified) type will be crucial for evaluation. The potential different GM-spruces may possess quite different inserted genes, but potentially several with an selective advantage.

Altered use

Some key questions about the effects of altered use of the organism, following genetic modification, are: (1) does area expansion occur, and (2) are new control agents favoured? Depending on the genes inserted different scenarios are possible e.g. use of GM-spruce by introduction to areas at present not regarded suitable for the species.

Area expansion
A number of otherwise unlikely environmental effects can occur. First, the environmental changes posed by the genetically modified organism will occur over a larger geographical area, and secondly, the fitness of the genetically modified organism becomes less important as a limiting factor for its spread, because this spread will be mediated through human action. Thus, altered use can increase the impacts on special species, habitats and ecosystems. In Norway spruce photoperiod is the environmental factor that initiates the cessation of growth and development of frost hardiness, but with some modifications caused by temperature.

Control agents
A number of transgenic plants which show resistance to a particular herbicide, are now produced. The use of these transgenic plants invites new usage of herbicides, which may be tougher – or more benign – to the general flora.

Transgenic Norway spruce

General considerations of invasion potential and gene flow to wild populations make it clear that Norway spruce is a high-risk GMO. Concerning GM-spruce the biodiversity dependent of the species Norway spruce must be included. In Figure 1 a subjective grading of the potential ecological and genetic risks is presented for a selected set of attributes of Norway spruce and the boreal (release) environment. Among these attributes, the most important ones for grading transgenic spruce at the high-risk end of the spectrum are (1) the essentially wild (self-propagating) characteristics of Norway spruce, (2) the broad geographic range it occupies, (3) the high level of gene flow among populations/neighbourhoods and the proximity of cultured (transgenic) stands to wild populations, (4) the important role that Norway spruce plays in the structure and function of boreal ecosystems, (5) the virtual impossibility of simulating realistic ecological conditions in the laboratory or in field test orchards, and (6) the typically uncontrolled access for the public to test sites. If transgenic spruce were to be based on only prereproductive (immature) stages outside of (controlled) greenhouses, the risks associated with reproduction and gene flow would decrease.

For some potential GM-types the main risk is invasion to new areas (e.g. frost tolerance) while other types affect the forest where spruce is the dominating species (e.g. enhanced pathogen resistance). Finally, both area types can be affected (e.g. enhanced drought resistance). A type of GM-spruce able to resist or change the decomposition of wood (e.g. insect resistance) is considered as specifically questionable due to threat to natural ecosystem function, habitats and species. The impact of using GM-spruce with enhanced fungi resistance is also uncertain, and can seriously affect mycorrhiza systems which are indispensable for ecosystem functioning.

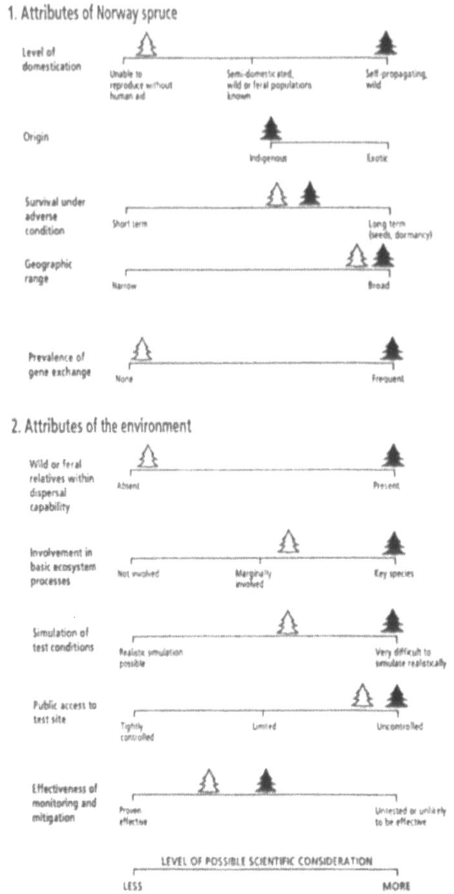

Figure 1. Subjective evaluation of the level of scientific considerations demanded by transgenic Norway spruce on a scale from less (low risk) to more (high risk). Two general situations are covered; one which involves reproductive stages (filled silhouettes) and one with obligate harvesting of immature trees (open silhouettes). From [1].

References

1. Tømmerås BÅ, Johnsen Ø, Skrøppa T, Hindar K, Holten J, Tufto J (1996) Long-term environmental impacts of release of transgenic Norway spruce (*Picea abies*). *NINA•NIKU Project Report* 003: 1–48
2. Kowarik I (1996) Auswirkungen von Neophyten auf Ökosysteme und deren Bewertung. *Texte des Umweltbundesamtes* 58/96 (Berlin), 119–155
3. Crawley MJ (1995) Long-term ecological impacts of the release of genetically modified organisms. *In*: *Pan-European conference on the potential long-term ecological impact of genetically modified organisms.* Council of Europe Press, Strasbourg, 43–66
4. Tiedje JM, Colwell RK, Grossman YL, Hodson RE, Lenski RE, Mack RN, Regal PJ (1989) The planned introduction of genetically engineered organisms: ecological considerations and recommendations. *Ecology* 70: 298–315
5. Simonsen L, Levin BR (1988) Evaluating the risk of releasing genetically engineered organisms. Special combined issue; *Trends Biotech* 6 *and Trends Ecol Evol* 3: 27–30
6. Drake JA, Mooney HA, di Castri F, Groves RH, Kruger FJ, Rejmanék M, Williamson M (eds) (1989) *Biological invasions: a global perspective. SCOPE* 37, John Wiley and Sons Ltd, Chichester, UK
7. Ryman N, Utter F, Hindar K (1995) Introgression, supportive breeding, and genetic conservation. *In*: JD Ballou, M Gilpin, T Foose (eds): *Population management for survival and recovery.* Columbia University Press, New York, 341–365
8. OECD (1986) *Recombinant DNA safety considerations.* Organisation for Economic Co-operation and Development, Paris
9. Royal Commission (1989) *The release of genetically engineered organisms to the environment.* Royal

Commission on Environmental Pollution, 13th report, HMSO, London
10. Griffith B, Scott JM, Carpenter JW, Reed C (1989) Translocation as a species conservation tool: status and strategy. *Science* 245: 477–480
11. Fremstad E, Elven R (1996) Alien plants in Norway. Sycamore (*Acer pseudoplatanus* L.). *Blyttia* 2: 61–78
12. Williamson M, Perrins J, Fitter A (1990) Releasing genetically engineered plants: present proposals and possible hazards. *Trends Ecol Evol* 5: 417–419
13. Ellstrand NC (1988) Pollen as a vehicle for the escape of engineered genes? Special combined issue; *Trends Biotech* 6 *and Trends Ecol Evol* 3: 30–32
14. Raybould AF, Gray AJ (1993) Genetically modified crops and hybridization with wild relatives: a UK perspective. *J Appl Ecol* 30: 199–219
15. Gliddon C (1994) The impact of hybrids between genetically modified crop plants and their related species: biological models and theoretical perspectives. *Mol Ecol* 3: 41–44
16. Waples RS (1991) *Definition of "species" under the Endangered Species Act: application to Pacific salmon.* NOAA Technical Memorandum, National Marine Fisheries Service, Seattle, F/NWC-194, 29 pp

Methods for Risk Assessment of Transgenic Plants
III. Ecologıcal rısks and propspects of transgenic plants
K Ammann, Y. Jacot, V. Simonsen and G. Kjellsson (eds)

Long-term questions related to agroecological effects of transgenic Bt-crops

Angelika Hilbeck and Franz Bigler

Swiss Federal Research Station for Agroecology and Agriculture, CH-8046 Zurich, Switzerland

Several genetically engineered crop plants (corn, cotton, potatoes) containing the gene from *Bacillus thuringiensis* (Berliner) (*Bt*) that encodes for the expression of an insecticidal δ-endotoxin have been commercially produced in the United States and many other countries since several years. The introduction of more and other transgenic *Bt*-crop plants to the US market and other countries is imminent [1, 2]. The safety of these transgenic *Bt*-plants released into the environment is currently of great concern to the public. One issue of the safety discussion is how transgenic *Bt*-plants will interact with other non-target organisms of different trophic levels.

Insecticides containing *Bt*-proteins have been used in agriculture for several decades. Based upon previous studies that were designed to test for undesired side-effects of *Bt*-insecticides on beneficial insects and the long record of safe use of commercial *Bt*-formulations, they are commonly considered to have little or no effect on natural enemies of pest insects [3–5]. However, *Bt*-insecticides and transgenic plants expressing *Bt*-proteins differ in a number of aspects. In currently commercially available transgenic plants, the *Bt*-proteins are produced in relatively high levels in almost all plant parts throughout the growing period until the plants senesce [6, 7]. Therefore, most if not all non-target herbivores colonizing transgenic *Bt*-plants that are not or sublethally affected to various degrees by the *Bt*-proteins will ingest plant tissue containing *Bt*-proteins which they may pass on to their natural enemies in a more or less processed form. Further, most *Bt*-proteins in transgenic plants are expressed in a truncated, activated form [6–8] that differs from the mixture of spores and crystalline, full-length *Bt*-proteins found in microbial *Bt*-insecticides [9]. *Bt*-corn plants produce a 69 kDa portion of the native, full-length 130 kDa Cry1Ab protoxin. This is a relatively small protoxin which is comprised of 620–648 amino acids [6]. Inside the insect gut only a small fragment must be further cleaved to produce the fully activated 65 kDa toxin. Consequently, no crystal solubilization and almost no protoxin-toxin conversion is necessary in the insect gut. But both processes are important for the specificity of *Bt*-compounds [10]. For example, the *in vitro* conversion of the 130 kDa Cry1Ac protoxin into the 60–65 kDa toxin involved seven specific cleavages, each removing fragments of about 10 kDa [11]. Thus, selectivity and biochemical processing may be altered in an insect.

Current assessment methods do not adequately account for the modified, activated form of release of *Bt*-proteins in transgenic plants and the extended duration of availability to herbivores and, consequently, also to other members of the food chain [12]. Therefore, appropriate assessment procedures have to be developed [13, 14].

As a first step towards this direction, we carried out tritrophic level model studies using transgenic *Bt*-corn plants (Cry1Ab) and the corresponding untransformed, *Bt*-free corn hybrid (both varieties were kindly provided by formerly Ciba Seeds) to study prey-mediated effects of *Bt*-

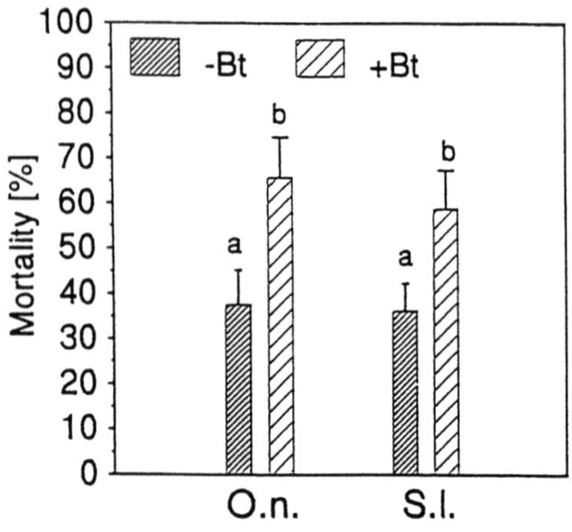

Figure 1. Total mean immature mortality [%] and standard error of *Chrysoperla carnea* larvae feeding on *Bacillus thuringiensis* corn-fed (*+Bt*) and *Bacillus thuringiensis*-free (*–Bt*) larvae of *Ostrinia nubilalis* (O.n.) and *Spodoptera littoralis* (S.l.) during their entire immature life stage (first instar to adult). (Columns with different letters represent treatment means that are significantly different at p = 0.05 (LSMEANS)).

corn on mortality of the predator *Chrysoperla carnea* (Stephens). Two different prey species were used in the experiments, the Egyptian cotton leafworm, *Spodoptera littoralis* (Boisduval), a lepidopterous, non-target pest and the European corn borer, *Ostrinia nubilalis* (Hübner), the lepidopterous, target pest. Another objective was to determine whether the observed effects were only due to intoxicated, suboptimal prey or could also be observed for non-target prey that was not or only slightly, sublethally affected. Small instar *S. littoralis* and *O. nubilalis* larvae were allowed to feed on the respective corn plant material for 12 to 24 h before they were provided as prey to *C. carnea* larvae. By then *O. nubilalis* larvae fed *Bt*-corn (*O. nubalis* (+)) exhibited early symptoms or disease but were not dying until another 24 to 48 h or more. *S. littoralis* larvae fed *Bt*-corn (*S. littoralis*(+)) did not exhibit any noticeable effects. Prey larvae were replaced every day by new prey that had fed for 12 to 24 h on the respective plant material. Total immature mortality for chrysopid larvae raised on *O. nubilalis*(+) or *S. littoralis*(+) was 59% and 66%, respectively, compared to 37% when raised on *Bt*-free prey of both species (Fig. 1) [15]. There was no significant difference in mortality between chrysopid larvae reared on *O. nubilalis*(+) or *S. littoralis*(+). Similarly, no significant difference in mortality was detected when chrysopid larvae were raised either on *O. nubilalis* (–) or *S. littoralis*(–). This suggested a direct *Bt*-induced effect.

To further investigate a direct *Bt*-induced effect, we fed the respective *Bt*-protein directly to *C. carnea* larvae using a novel bioassay technique which allowed for incorporation of the activated Cry1Ab toxin into a liquid diet specifically developed for optimal nutrition of *C. carnea*. This media was then encapsulated within small paraffin spheres. Because only second and third instars can penetrate the skins of the paraffin spheres, two different methods were used to rear chrysopid larvae through first instar. The first method used 0.5 cm^3 foam cubes soaked in non-encapsulated, liquid diet. For one treatment, activated Cry1Ab toxin (100 µg/ml diet) was mixed into the non-encapsulated diet whereas only an equivalent amount of double-distillated water

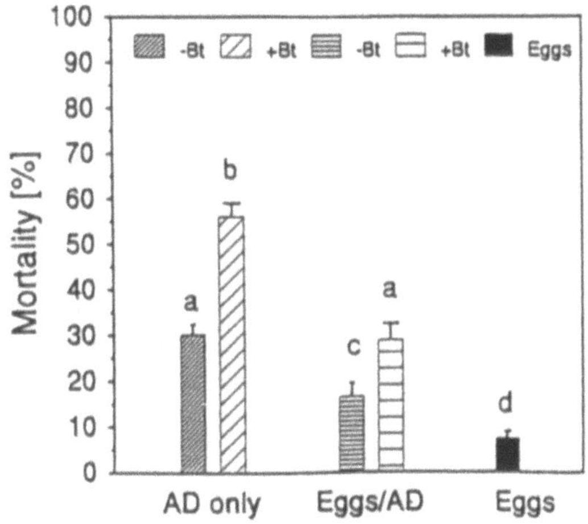

Figure 2. Total mean immature mortality [%] and standard error of *Chrysoperla carnea* larvae feeding on different types of Cry1Ab toxin-containing and untreated diets during their entire immature life stage (first instar to adult). (Means with different letters are significantly different at p = 0.05 significance level (LSMEANS); AD = artificial diet only including first instar; Eggs/AD = *E. kuehniella* eggs during first instar, artificial diet during second and third instar; Eggs = *E. kuehniella* eggs only)

was added to the diet for the corresponding control. The second method used *Ephestia kuehniella* (Hübner) eggs as prey during first instar. After reaching second instar, all larvae received encapsulated, artificial diet with or without Cry1Ab, respectively. In a fifth treatment, chrysopid larvae were raised on *E. kuehniella* eggs only. When reared only on artificial diet containing Cry1Ab toxin, total immature mortality was significantly higher (56%) than in the respective untreated control (30%) (Fig. 3). Also, significantly more chrysopid larvae died (29%) that received Cry1Ab later during their larval development compared to the respective control (17%). Only 8% of the larvae died when reared exclusively on *E. kuehniella* eggs (Fig. 2). These results demonstrate that activated Cry1Ab is toxic to *C. carnea* at 100 µg/ml diet. However, mortality in these direct-feeding studies was similar to the mortality detected in the tritrophic studies using transgenic *Bt*-plants despite the large difference in *Bt*-concentrations between the artificial diet (100 µg/ml diet) and those expressed in transgenic *Bt*-plants (2 13–3.27 µg/g fresh weight [16]). This suggests that interactions between the herbivores and *Bt*-plants occur that either cause unnoticed secondary effects in the herbivore or further process the *Bt*-protein rendering the prey (+) more toxic to *C. carnea.*

These findings illustrate that different approaches are necessary to determine reliably the long-term agroecological consequences of transgenic *Bt*-plants. Field studies have to be conducted to determine the ecological consequences of these laboratory results. Monitoring programs should be established that assess the long-term compatibility of naturally-occurring biological control with the utilization of transgenic *Bt*-crop plants to secure the sustainable use of transgenic *Bt*-crop plants that were introduced as a significant contribution to a more environmentally friendly agriculture. Performance and fitness of natural enemies in *Bt*-crop fields also may affect pest resistance development [17, 18].

Further, we want to emphasize that for an overall ecological risk assessment of transgenic *Bt*-plants, other risk aspects must be included and analyzed in context with each other. Aside of intensively studied pest resistance development, these include the effects of *Bt*-plant residues on non-target organisms in soils. Only little has been done in this field of research until today, despite the fact that researchers at the New York University have found lab-produced *Bt*-proteins to be persistent and retain insecticidal activity for many weeks in certain soil types [19, 20]. Further, effects of *Bt*-proteins on non-target herbivores in the agroecosystem have to be studied because they can potentially change prey diversity and quantity for natural enemies thereby affecting natural enemy – prey relationships [21]. Miller [21] found for example a significant reduction of species richness and densities of non-target lepidopteran herbivores when *Bt*-insecticides were applied in a gypsy moth (*Lymantria dispar*) eradication program in Oregon forests from 1986 to 1988. Finally, also outcrossing of *Bt*-genes to wild/weedy relatives needs to be evaluated carefully. If *Bt*-genes escape to relatives in natural habitats, possible effects of *Bt*-proteins described for agroecosystems also apply in natural habitats and the interface of agricultural and natural habitats, respectively (e.g. alterations in arthropod communities and food webs), which may in turn again affect processes in agricultural habitats.

References

1. Hoyle R (1995) EPA okays first pesticidal transgenic plants. *Bio/Technology* 13: 434–435
2. Niebling K (1995) Agricultural biotechnology companies set their sights on multi-billion $ markets. *Genetic Engineering News* 15: 20–21
3. Croft BA (1990) *Arthropod biological control agents and pesticides*. John Wiley and Sons, New York, 723 pp
4. Flexner JL, Lighthart B, Croft BA (1986) The effects of microbial pesticides on non-target, beneficial arthropods. *Agr Ecosyst Environ* 16: 203–254
5. Melin BE, Cozzi EM (1989) Safety to nontarget invertebrates of lepidopteran strains of *Bacillus thuringiensis* and their β-exotoxins. *In*: M Laird, LA Lacey, EW Davidson (eds): *Safety of microbial insecticides*. CRC Press, Boca Raton, Florida
6. Koziel M G, Beland GL, Bowman C, Carozzi NB, Crenshaw R, Crossland L, Dawson J, Desai N, Hill M, Kadwell S, Launis K, Lewis K et al.. (1993) Field performance of elite transgenic maize plants expressing an insecticidal protein derived from *Bacillus thuringiensis*. *Bio/Technology* 11: 194–200
7. Perlak FJ, Deaton RW, Armstrong TA, Fuchs RL, Sims SR, Greenplate JT, Fischhoff DA (1990) Insect resistant cotton plants. *Bio/Technoloy* 8: 939–943
8. Fujimoto H, Itoh K, Yamamoto M, Kyozuka J, Shimamoto K (1993) Insect resistant rice generated by introduction of a modified δ-endotoxin gene of *Bacillus thuringiensis*. *Bio/Technology* 11: 1151–1155
9. Feitelson JS, Payne J, Kim L (1992) *Bacillus thuringiensis*: Insects and Beyond. *Bio/Technology* 10: 271–275
10. Visser B, Bosch D, Honée G (1993) Domain-function studies of *Bacillus thuringiensis* crystal proteins: A genetic approach. *In*: PF Entwistle, JS Cory, MJ Bailey, S Higgs (eds): *Bacillus thuringiensis, an environmental biopesticide: Theory and practice*. John Wiley and Sons, Chichester, UK
11. Choma CT, Surewicz WK, Carey PR, Pozsgay M, Raynor R, Kaplan H (1990) Unusual proteolysis of the protoxin and toxin from *Bacillus thuringiensis*. Structual implications *Eur J Biochem* 189: 523–527
12. Jepson PC, Croft BA, Pratt GE (1994) Test systems to determine the ecological risks posed by toxin release from *Bacillus thuringiensis* genes in crop plants. *Molec Ecol* 3: 81–89
13. Goldburg RJ, Tjaden G (1990) Are B.T.K. plants really safe to eat? *Bio/Technology* 8: 1011–1015
14. Snow AA, Palma PM (1997) Commercialization of transgenic plants: Potential ecological risks. *BioScience* 47: 86–96
15. Hilbeck A, Baumgartner M, Fried PM, Bigler F (1998) Effects of transgenic *Bacillus thuringiensis*-corn-fed prey on mortality and development time of immature *Chrysoperla carnea* (Neuroptera: Chrysopidae). *Environ Entomol* 27: 480–487
16. Fearing PL, Brown D, Vlachos D, Meghji M, Privalle L (1997) Quantitative analysis of CryIA(b) expression in *Bt* maize plants, tissues, and silage and stability of expression over successive generations. *Mol Breeding* 3: 169–176
17. Gould F, Kennedy GG, Johnson MT (1991) Effects of natural enemies on the rate of herbivore adaptation to resistant host plants. *Entomol Exp Appl* 58: 1–14
18. Johnson MT, Gould F, Kennedy GG (1997) Effects of natural enemies on relative fitness of *Heliothis virescens* genotypes adapted and not adapted to resistant host plants. *Entomol Exp Appl* 82: 219–230
19. Tapp H, Stotzky G (1995) Dot blot enzyme-linked immunosorbent assay for monitoring the fate of insecticidal toxins from *Bacillus thuringiensis* in soil. *Appl Environ Microbiol* 61: 602–609

20. Koskella J, Stotzky G (1997) Microbial utilization of free and clay-bound insecticidal toxins from *Bacillus thuringiensis* and their retention of insecticidal activity after incubation with microbes. *Appl Environ Microbiol* 63: 3561–3568
21. Miller JC (1990) Field assessment of the effects of a microbial pest control agent on nontarget lepidoptera. *Amer Entomol* 36: 135–139

Discussion session 3: Short-term, long-term effects and standardization of limits

Phil Dale
Someone pointed out that we should not focus that much on gene dispersal because we have quite a lot of experience and data on this, and now it is important that we focus on effects and look at the specific race, specific genes. The choice of efficient methods is difficult, but you could build on experience from toxicology, you have a lot of things you can test there in soil organisms, insects and so on. We need to define which ecosystems we are going to use for these studies.

The quantification of test data to rely on is then next most important besides the experience. We can also use of the mass of the experience from agricultural tradition. Criteria for rejection and acceptance have to be defined. That is something we must discuss in relation to possible effects. It is a very complex area.

Jan Husby
What should we do about risk assessment for trees in general? They have a very long generation time. That is a specific, a difficult situation in risk assessment.

What crucial things do we need to know before approving the release of genetically modified plants?

Bjørn Tommerås
In risk assessments there are "need to knows" and "nice to knows". And there are a lot of things that will be nice to know. The main impact of a herbicide tolerance is going to be principally an agronomic one. And it is really a matter of how we develop a agriculture practice. The main questions are: Can we design ways of managing three different herbicide tolerant genes together (referring to Xavier Reboud's talk)? Will it affect control of volunteers? Will it affect control in the next crop culture that also contains one of those herbicide tolerant genes? So in the sense of herbicide tolerance, I am sure there will be people here that will ask questions about the impact on natural habitats also. But the answers we need are in relation to how we managed the transgenic plant culture in practice. The extensive transgenic colza cultures in Canada have to be closely watched, but you cannot compare them directly to Europe with its different agriculture practice. So some of the questions and answers will have to be different. In a way, we can only do that really effectively by commercializing. There are other issues where we need to have a degree of confidence before we can commercialize, but we really need to commercialize in a large scale and find out what the problems are in practice. We have to define what monitor questions we are going to ask and get that data back and if after some years, if it looks satisfactory we have strategies to cope with.

Lisa Rudenko
I want to bring up a very important point that risk assessment is an iterative process. We are not simply identifying a hazard by asking: can this occur? We should ask: does it occur, and under what circumstances and at what rate?

Klaus Ammann
The big biotech companies are highly interested to talk about these matters and to go to a more sustainable agriculture. Another point is that we should also carefully screen what the biotech companies have already done. And with the *Bt* field-studies a lot of work has been done, maybe not enough, but there are studies showing that the pesticide application is much more harmful to the beneficial insects than the *Bt*-strategy. Several papers state that and we have to consider those papers and to carefully listen to each other. I agree with Angelika Hilbeck on the fact that we need more studies and more specific studies outside in the field, because containment is often producing a very harsh selection environment for the beneficial insects: It has nothing to do with reality when those poor lacewings have to feed on paraffin spheres coated with purified Bt protein.

We also have to remenber when saying that there are always going to be impacts, that impacts should be just taken both positively and negatively. And that there is a value judgment done on friendly and unfriendly. The regulators try to balance this.

Piet Schenkelaars
How and when are you as a regulator able to make a decision in a situation when the research is still going on about many ecological effects?

Thomas Nickson
Speaking as a scientist, I do not look at the regulatory approval as a "drop dead endpoint". It is an endpoint from which we can proceed to the next level. We proceeded to commercialization, but that does not mean that any product is beyond the scrutiny of stewardship, the scrutiny that the company gives their products in order to make sure there are continuing to be safe and continue to be everything that they should be for the consumer and in this case it is the farmer and that is a critical aspect. So we continue to evaluate our products to ensure safety and the highest possible quality for the consumer.

Philippe Gay
As a regulator, I see the safety analysis process as a dynamic process that is never finished. For me, the main question is, when do I have to come back to the decision? This means that regulators are confronted with the question of when new data would be the cause of changing approvals?

Session 4: Monitoring methods

Molecular markers for monitoring transgenic plants

Vibeke Simonsen

Department of Terrestrial Ecology, National Environmental Research Institute, Vejlsoevej 25, P.O. Box 314, DK-8600 Silkeborg, Denmark

Introduction

Monitoring is a continuous control of something. In population biology it can be the study of populations during a time period, e.g. the study of succession in an ecosystem with the focus on a certain species or genotype, or dispersal of a species in a certain area. Depending on the species studied the questions of why, when and how the monitoring shall be performed have to be considered before initiating the monitoring program. Marvier et al. [1] have discussed if monitoring is the way to reduce the risks of transgenic plants.

Transgenic plants offer an excellent possibility for monitoring dispersal of specific genes as the plants in their genome contain a well-characterized transgene. Until now various transgenes have been used [2], which can be classified in four groups [3]: Crop protection against diseases, herbicide tolerance, stress tolerance and quality traits, encompassing natural and alien plant products, e.g. pharmaceuticals. Several genes are used for crop protection against diseases, and especially genes providing resistance to insects have been applied. Tobacco plants, producing proteinase inhibitor, are shown to have an impact on the species diversity of degrading organisms, living in soil [4]. Genes from the bacterium *Bacillus thuringiensis* coding for insecticidal crystal proteins, which are not toxic to humans and many other non target species, have obtained a great application as crop protection genes. However, the target insect species can develop a resistance to the specific toxin produced by the plant, which may be reversed, when removing the selection pressure [5], i.e. when plants are not producing the toxin. These observations imply that many of the topics studied in population biology are relevant for studying the performance of transgenic plants. Herbicide tolerance as well as stress tolerance may cause severe problems to agriculture if the transgenes are dispersed either as cultivars outside the fields or transferred to related natural species or genera, which are considered to be weeds. Transmission of a gene or a transgene can occur as found for, for example *Brassica* [6–10], *Beta* [11] and *Sorghum* [12], and hence the possibility to obtain herbicide tolerant weed or weed with increased stress tolerance is present. In both cases regional differences have to be considered, e.g. transgenic potato has been claimed to be of no risk for Canada and United States [13], but for the original source of potato, the Andes region, the risk of getting unwanted transmission of transgenes is possible. Changed quality traits may also have an impact on the composition of metabolites which again may change the food quality of the plants for herbivores. The implication of the production of pharmaceuticals by transgenic plants has to be studied carefully to avoid unwanted effects on the ecosystem.

Transgenic plants are, as mentioned in the previous paragraph, very well suited for monitoring programs with the aim to look for dispersal of a transgene as the transgene can serve as a molecular marker. Molecular markers encompass nucleic acids, proteins and metabolites. A depiction of the relation among molecular markers is shown in Figure 1, which is divided into a group of markers providing the genetic information and one on the expression of the genetic information. It has to be emphasized that molecular markers as DNA may encompass genes as well as nonsense DNA sequences. Molecular markers are widely applied for the study of populations [e.g. 14, 15], and also as versatile tools in breeding programs [16]. The molecular markers are inherited according to the Mendelian laws, and they can have dominant, codominant or recessive inheritance.

DNA (DNA organized in chromosomes) Information
↓
RNA Expression
↓
Protein (*e.g.* enzyme)
↓
substrate → product (*e.g.* metabolite)

Figure 1. Molecular markers.

Molecular markers are used for individual identification as well as parentage identification in forensic medicine and for grouping of individuals in population biology. The mode of grouping can be helpful in understanding mating structure, migration among populations, selection and other features important for the dynamics of the population. Furthermore, phylogenies may be constructed on data from molecular markers. Statistical analyses as well as program packages for analysing data obtained from studies of molecular markers in populations have been developed [e.g. 17–20].

The aim of this presentation is to give a brief summary on the methods which can be used for detection of molecular markers and with the emphasis on methods relevant for studying plants, containing a transgene.

DNA

An absolute necessary assumption for studying molecular markers is the presence of a sequence of DNA, which may code for a gene or not. Also it has to be mentioned that a gene may be present in the genome but not expressed, which may be due to age dependent or tissue dependent expression of the gene. Many transgenic constructs consist of a constitutive promoter, which ensures the expression of a specific gene, independent of age and tissue, a gene needed for the wanted product and maybe a marker gene (see the following section on marker genes). The methods for detecting the presence of a gene in the genome are developing very rapidly and have been improved tremendously with regard to the needed time and cost per analyzed sample. The methods can be divided into two groups, one using DNA directly isolated, and the other using amplification of DNA by performing a polymerase chain reaction (PCR) [21]. Moreover, the DNA may be treated with restriction enzymes, which cuts the DNA at specific sites. The method using isolated DNA, digestion with restriction enzymes, electrophoresis of the fragments, transfer of the fragments to a membrane and visualization by applying the appropriate probe, marked radioactively or biotinylated, is called restriction fragment length polymorphism (RFLP), and the visualization process is a Southern blot [22]. The advantage of transgenic plants is the knowledge of the alien DNA which facilitates design of probes for identification of the plants carrying the construct [23]. The RFLP method has also been applied to bulk analysis in order to detect different DNA fragments encoding wanted traits, the result of the procedure is identification of a quantitative trait locus (QTL) [e.g. 24]. In a similar way as RFLP uses gene characteristic primers in the PCR process, amplification of the gene of interest can be performed, and the product can be visualized with ethidiumbromide or other dyes after an electrophoretic separation. By applying any of these methods the presence of the transgenic construct can be proven with a high probability, but transcription of the construct is not verified. For this, detection methods as described in the next sections on proteins and metabolites, respectively, are needed. Single sequence repeats (SSR), which are repeats of a few bases (e.g. AATAATAAT...), are often highly variable in number in a population. When applying PCR amplification with specific primers for the ends of the repeats the sequence can be amplified, and the product detected after electrophoresis. This process can be automated and the number of alleles, when applying the method for determination of SSR, may be very great and therefore well suited for identification of hybrids, dispersal of genes etc. [e.g. 25]. However, if this method shall be used for detection of transgenic plants, a specific allele shall be present in the plant material transformed and strongly associated to the transgenic construct. A recent method, called amplified fragment length polymorphism (AFLP), combines the digestion with restriction enzymes, amplification by the PCR reaction, electrophoresis and visualization [e.g. 26–28]. This method reveals a huge amount of variation, but for the study of transgenic plants the use of restriction enzymes and Southern blot are easier to perform and will provide the result, detection of the transgenic construct. Several of the methods can be used directly for detection of a specific gene [e.g. 29] and others for identification of hybrids [e.g. 30]. The DNA methods by using gene-specific primers or probes have provided a great step further in determination of linkage groups on the chromosomes and these methods as such have been reviewed [31]. These methods may be useful for testing the presence and the localization of a gene and may also be adequate for monitoring of transgenic plants.

RNA

RNA is necessary for the transmission of genetic information from DNA to protein, as depicted in Figure 1. The presence of a certain RNA indicates that the organism has the ability to produce the corresponding gene product. RNA for specific genes can be detected by probes which can be marked with either radioactivity or bio-luminescence (Northern blot). Often the amount of RNA is very limited so a PCR amplification is necessary [32]. Electrophoresis can be used for separation of the RNA fragments and these can be visualized by a Northern blot when transferring the separated RNA fragments to a membrane and applying the appropriate probe to the membrane. The method has been widely applied for detecting the presence of a transgene [e.g. 33–36]. Methods for quantification of the RNA products are also available [37, 38], which may be of interest for detecting how much energy the plant is devoting for the production of the particular product. However, all the methods mentioned are only suited for monitoring transgenic plants in small scale experiments.

Proteins

The methods for detection of proteins may rely either on the composition of amino acids of the molecule, the electric charge, the size, the enzymatic or the immunological ability of a protein. The method for determination of the amino acid composition, amino acid sequencing, is, despite the fact that the procedure can be highly automatic, a time-consuming process. It is needed to isolate the protein in question before the analysis can take place and the procedure demands advanced equipment, so the method may not be realistic for monitoring transgenic plants. The utilization of the electric change of a protein is done when performing gel electrophoresis. The method separates a mixture of proteins according to their individual electric charge in a gel, made from, for example, agarose, starch, cellulose acetate or polyacrylamide. Depending of the material used for the gel matrix also the molecular size can have an effect on the separation of the proteins. This method can be combined with a visualization procedure for proteins as such or utilizating the enzymatic or immunological ability of the protein; further information on gel electrophoresis is available in [39, 40]. Descriptions of procedures for enzymatic visualization have been compiled [e.g. 41]. When using the immunological ability the relevant antibody has to be produced and information on how this is done as well as further description of the technical details is available [42, 43]. It has to be emphasized that the production of the particular antibody is the greatest task when dealing with immunological methods. Recently, methods for detecting expression of a certain protein in various tissues have been developed, e.g. immunogold labelling, where gold particles are coated with the relevant antibody, applied to the tissue slices and scanned with an electronic microscope. Methods utilizing the immunological property of the protein seem to be used increasingly also in the context of transgenic plants for detecting tissue differential expression of a gene product [e.g. 44–46]. The advantage of using enzymatic or immunological property of a protein in a mixture of proteins is that a great number of individuals can be checked quickly, especially when using the immunological methods as enzyme linked immunosorbent assay (ELISA) or radioimmunoassay (RIA).

Metabolites

Metabolites are in this case considered to be non-protein molecules, and rather, organic compounds. These compounds can be defence substances such as proline, cyanide, phenolic compounds, etc., or changed quality traits such as oil, starch, sugar etc. Several chemical analytic methods are available depending on the structure of the chemical compound. Among these are chromatographic methods, which utilize the solubility of the compound in various phases. The relative simple method thin-layer chromatography (TLC) consists of a stationary phase, e.g. silica gel coated on a glass plate or plastic sheet, and a mobile phase containing a solvent. The samples are applied on the stationary phase. The plate or sheet is placed in a solvent reservoir and due to capillary forces the solvent migrates upward in the stationary phase. The samples will then be separated according to their solubility in the stationary and mobile phase, and when the solvent passes the other edge of plate, the analysis is stopped and a visualization procedure for the chemical compound performed. High performance liquid chromatography (HPLC) is a separation of chemical compounds in a mobile liquid phase and a stationary phase, consisting of a liquid on a solid support or a solid, under high pressure, and the compounds are detected as a change in refractive index, light absorption or fluorescence. In gaschromatography (GC) the mobile phase is a gas, and specific detectors are developed. Capillary electrophoresis is performed in capillaries under high voltage, and detected in a similar way as used for HPLC. Advanced methods such as HPLC, GC or capillary electrophoresis need advanced equipment, skilled persons for doing the analysis and have compared to TLC a limited capacity regarding the number of samples which can be analysed per day. TLC has been applied for detecting hybrids [e.g. 47], so the method may be possibly applied for transgenic plants producing a specific chemical compound.

Marker genes

A shortcut for rapid detection of transgenic plants is the use of marker genes. The marker gene is attached to the transgenic construct. Neomycin phosphotransferase II and β-glucuronidase have both been used as markers for transgenic constructs. Neomycin phosphotransferase II provides resistance to kanamycin, which can be tested by a bio-assay, whereas presence of β-glucuronidase can be detected by a histochemical analysis, which now can be performed as a chemolumiscence assay [48]. Marker genes, providing resistance, may cause an enhanced risk for dispersal of unwanted resistant genes, e.g. to human pathogens, so it ought to be avoided for future transgenic constructs. Luciferase, which can be detected by embedding the tissue in a solution of luciferin and adenosine triphosphate (ATP) and then exposed to light [49, 50], and green fluorescent protein from jellyfish [51–53], which is visible when exposed to ultra violet light (UV-light), have both been suggested for this purpose. Neither luciferase nor green fluorescent protein so far have shown unwanted effects. However, the use of markers genes for identification of transgenic plants may not be unambiguous, so additional investigation for identification of the gene, producing the wanted product, or detection of the product, may be needed.

Demand for monitoring methods

The monitoring methods shall be easy to perform for a huge number of plants. In summary, the following demands shall at least be fulfilled: High reliability, inexpensive, non- or minor destructive, and results in the form of presence/absence. Many of the methods mentioned in the previous sections are complicated, need expensive equipment and skilled personnel to perform. Simple techniques like bio-assays or dot-tests which may be performed either in microtiter plates or on membranes, where the application of the necessary solutions can be done automatically, and which only need an application of a small piece of leaf, so the plant can continue growth, may be useful. In this context the knowledge of the transgenic construct can be of great help for designing appropriate probes.

In conclusion it can said that there is a need for development of simple monitoring methods suited for transgenic plants, if society wants to keep track of the dispersal of transgenic plants and transgenic constructs, but there is also a need for knowledge on why, when and how the monitoring shall be performed [51].

Acknowledgement
Dr. R. Bagger Jørgensen is acknowledged for valuable comments on an earlier version of this manuscript.

References

1. Marvier M, Meir E, Kareiva P (1998) Is monitoring the way to reduce the risks of genetically engineered crops? This volume, 109–122
2. Madsen KH, Poulsen GS (1997) Inserted traits for transgenic plants. *In*: G Kjellsson, V Simonsen, K Ammann (eds): *Methods for risk assessment of transgenic plants. II Pollination, gene-transfer and population impacts.* Birkhäuser, Basel, 203–219
3. Rogers HJ, Parkes HC (1995) Transgenic plants and the environment. *J Exp Bot* 46: 467–488
4. Donegan KK, Seidler RJ, Fieland VJ, Schaller DL, Palm CJ, Ganio LM, Cardwell DM, Steinberger Y (1997) Decomposition of genetically engineered tobacco under field conditions: Persistence of the proteinase inhibitor I product and effects of soil microbial respiration and protozoa, nematode and microarthropod populations. *J Appl Ecol* 34: 767–777
5. Tabashnik BE, Finson N, Groeters FR, Moar WJ, Johnson MW, Luo K, Adand MJ (1994) Reversal of resistance to *Bacillus thuringiensis* in *Plutella xylostella. Proc Natl Acad Sci USA* 91: 4120–4124
6. Mikkelsen TR, Andersen B, Jørgensen RB (1996) The risk of crop transgene spread. *Nature* 380: 31
7. Frello S, Hansen KR, Jensen J, Jørgensen RB (1995) Inheritance of rapeseed (*Brassica napus*)-specific RAPD markers and a transgene in the cross *B. juncea* × (*B. juncea* × *B. napus*). *Theor Appl Genet* 91: 236–241
8. Jørgensen RB, Andersen B (1994) Spontaneous hybridization between oilseed rape (*Brassica napus*) and the weedy *B. campestris* (Brassicaceae): A risk of growing genetically modified oilseed rape. *Amer J Bot* 81: 1620–1626
9. Lefol E, Danielou V, Darmency H, Boucher F, Maillet J, Renard M (1995) Gene dispersal from transgenic crops. I. Growth of interspecific hybrids between oilseed rape and the wild hoary mustard. *J Appl Ecol* 32: 803–808
10. Chèvre AM, Eber F, Baranger A, Renard M (1997) Gene flow from transgenic crops. *Nature* 389: 924
11. Bartsch D, Pohl-Orf M (1996) Ecological aspects of transgenic sugar beet: Transfer and expression of herbicide resistance in hybrids with wild beets. *Euphytica* 91: 55–58
12. Arriola PE, Ellstrand NC (1996) Crop-to-weed gene flow in the genus *Sorghum* (Poaceae): Spontaneous interspecific hybridization between johnsongrass, *Sorghum halepense*, and crop sorghum, *S. bicolor. Amer J Bot* 83: 1153–1160
13. Love SL (1994) Ecological risk of growing potatoes in the United States and Canada. *Amer Potato J* 71: 647–658
14. Soltis PS, Soltis DE (1991) *Isozymes in plant biology*. Chapman and Hall, London
15. Powers DA (1993) Application of molecular techniques to large marine ecosystems. *In*: K Sherman, L Alexander, B Gold (eds): *Large marine ecosystems V: Stress, mitigation and sustainability of large marine ecosystems.* American Association for the Advancement of Science, Washington, 320–352
16. Quarrie SA (1996) New molecular tools to improve the efficiency of breeding for increased drought resistance. *Plant Growth Regul* 20: 167–178
17. Richardson BJ, Baverstock PR, Adams M (1986) *Allozyme electrophoresis*. Academic Press, San Diego
18. Weir BS (1990) Intraspecific differentiation. *In*: DM Hillis, C Moritz (eds): *Molecular systematics*. Sinauer Associates, Sunderland, 373–410
19. Swofford DL, Olsen GJ (1990) Phylogeny reconstruction. *In*: DM Hillis, C Moritz (eds): *Molecular systematics*. Sinauer Associates, Sunderland, 411–501
20. Shierup MH (1997) M8. Computer programs for analysis of genetic data. *In*: G Kjellsson, V Simonsen, K Ammann (eds): *Methods for risk assessment of transgenic plants. II Pollination, gene-transfer and population impacts.* Birkhäuser, Basel, 95–96
21. Innis MA, Gelfand DH (1990) Optimization of PCRs. *In*: MA Innis, DH Gelfand, JJ Sninsky, TJ White (eds): *PCR protocols: A guide to methods and applications*. Academic Press, San Diego, 3–12
22. Dowling TE, Moritz C, Palmer JD (1990) Nucleic acids II: Restriction site analysis. *In*: DM Hillis, C Moritz (eds): *Molecular systematics*. Sinauer Associates, Sunderland, 250–317
23. Register III, JC (1997) Approaches to evaluating the transgenic status of transformed plants. *Trends Biotech* 15: 141–146
24. Stuber CW (1990) Molecular markers in the manipulation of quantitative characters. *In*: AHD Brown, MT Clegg, AL Kahler, BS Weir (eds): *Plant population: Genetics, breeding, and genetic resources*. Sinauer

Associates, Sunderland, 334–350
25. Powell W, Morgante M, McDevitt R, Vendramin GG, Rafalski JA (1995) Polymorphic simple sequence repeat regions in chloroplast genomes: Applications to the population genetics of pines. *Proc Natl Acad Sci USA* 92: 7759–7763
26. Vos P, Hogers R, Bleeker M, Reijans M, van de Lee T, Hornes M, Frijtes A, Pot J, Peleman J, Kuiper M, Zabeau M (1995) AFLP: A new technique for DNA fingerprinting. *Nucl Acid Res* 23: 4407–4414
27. Becker J, Vos P, Kuiper M, Salamini F, Heun M (1995) Combined mapping of AFLP and RLFP markers in barley. *Mol Gen Genet* 249: 65–73
28. Thomas CM, Vos P, Zabeau M, Jones DA, Norcott KA, Chadwick BP, Jones JDG (1995) Identification of amplified restriction fragment polymorphism (AFLP) markers tightly linked to the tomato Cf-9 gene for resistance to *Cladosporium fulvum*. *Plant J* 8: 785–794
29. Scheffler JA, Parkinson R, Dale PJ (1995) Evaluating the effectiveness of isolation distances for field plots of oilseed rape (*Brassica napus*) using a herbicide-resistance transgene as selectable marker. *Plant Breeding* 114: 317–321
30. D'Hont A, Rao PS, Feldmann P, Grivet L, Islam-Faridi N, Taylor P, Glaszmann JC (1995) Identification and characterisation of sugarcane intergeneric hybrids, *Saccharum officinarum* × *Erianthus arundinaceus*, with molecular markers and DNA *in situ* hybridization. *Theor Appl Genet* 91: 320–326
31. Jiang J, Gill BS (1994) Nonisotopic *in situ* hybridization and plant genome mapping: The first 10 years. *Genome* 37: 717–725
32. Kawasaki ES (1990) Amplification of RNA. *In*: MA Innis, DH Gelfand, JJ Sninsky, TJ White (eds): *PCR protocols: A guide to methods and applications*. Academic Press, San Diego, 21–27
33. Hsieh H-L, Tong C-G, Thomas C, Roux SJ (1996) Light-modulated abundance of an mRNA encoding a calmodulin-regulated chromatin-associated NTPase in pea. *Plant Mol Biol* 30: 135–147
34. Kato N, Esaka M (1996) cDNA cloning and gene expression of ascorbate oxidase in tobacco. *Plant Mol Biol* 30: 833–837
35. Torelli A, Soragni E, Bolchi A, Petrucco S, Ottonello S, Branca C (1996) New potential markers of *in vitro* tomato morphogenesis identified by mRNA differential display. *Plant Mol Biol* 32: 891–900
36. Vancanneyt G, Scmidt R, O'Connor-Sanchez A, Willmitzer L, Rocha-Sosa M (1990) Construction of an intron-containing marker gene: Splicing of the intron in transgenic plants and its use in monitoring early events in *Agrobacterium*-mediated plant transformation. *Mol Gen Genet* 220: 245–250
37. Gilliland G, Perrin S, Bunn HF (1990) Competitive PCR for quantitation of mRNA. *In*: MA Innis, DH Gelfand, JJ Sninsky, TJ White (eds): *PCR protocols: A guide to methods and applications*. Academic Press, San Diego, 60–69
38. Wang AM, Mark DF (1990) Quantitative PCR. *In*: MA Innis, DH Gelfand, JJ Sninsky, TJ White (eds): *PCR protocols: A guide to methods and applications*. Academic Press, San Diego, 70–75
39. Murphy RW, Sites Jr JW, Buth DC, Haufler CH (1990) Proteins I: Isozyme electrophoresis. *In*: DM Hillis, C Moritz (eds): *Molecular systematics*. Sinauer Associates, Sunderland, 45–126
40. Westermeier R (ed) (1993) *Electrophoresis in practice: A guide to theory and practice*. VCH Verlagsgesellschaft mbH, Weinheim
41. Manchenko GP (1994) *Handbook of detection of enzymes on electrophoretic gels*. CRC Press, Boca Raton
42. Axelsen NH, Krøll J, Weeke B (eds) (1973) *A manual of quantitative immunoelectrophoresis: Methods and applications*. Universitetsforlaget, Oslo
43. Maxson LR, Maxson RD (1990) Proteins II. Immunological techniques. *In*: DM Hillis, C Moritz (eds): *Molecular systematics*. Sinauer Associates, Sunderland, 127–155
44. Hippe S, Düring K, Kreuzaler F (1989) *In situ* localization of a foreign protein plants by immunoelectron microscopy following high pressure freezing. Freeze substitution and low temperature embedding. *Eur J Cell Biol* 50: 230–234
45. Manteuffel R, Panitz R (1993) *In situ* localization of faba bean and oat legumin-type proteins in transgenic tobacco seeds by a highly sensitive immunological tissue print techniques. *Plant Mol Biol* 22: 1129–1134
46. Benhamou N (1995) Immunocytochemistry of plant defence mechanisms induced upon microbial attack. *Microsc Res Techn* 31: 63–73
47. Veit M, Bauer K, Beckert C, Kast B, Geiger H, Czygan FC (1995) Phenolic characters of British hybrid taxa in *Equisetum* subgenus Equisetum. *Biochem Syst Ecol* 23: 79–87
48. Information available on http://www.tropix.com
49. Chia T-F, Chan Y-S, Chua N-H (1994) The firefly luciferase gene as a non-invasive reporter for *Dendrobium* transformation. *Plant J* 6: 441–446
50. Kost B, Schnorf M, Potrykos I, Neuhaus G (1995) Non-destructive detection of firefly luciferase (LUC) activity in single plant cells using a cooled, slow scan CCD camera and an optimized assay. *Plant J* 8: 155–166
51. Sheen J, Hwang S, Niwa Y, Kobayshi H, Galbraith DW (1995) Green-fluorescent protein as a new vital marker in plant cells. *Plant J* 8: 777–784
52. Stewart CN (1996) Monitoring transgenic plants using *in vivo* markers. *Nat Biotechnol* 14: 682
53. Haseloff J, Siemering KR, Prasher DC, Hodge S (1997) Removal of a cryptic intron and subcellular localization of green fluorescent protein are required to mark transgenic *Arabidopsis* plants brightly. *Proc Natl Acad Sci USA* 94: 2122–2127

Methods for Risk Assessment of Transgenic Plants
III. Ecological risks and propspects of transgenic plants
K. Ammann, Y. Jacot, V. Simonsen and G. Kjellsson (eds)

Biogeographical assay and natural gene flow

Pia Rufener Al Mazyad and Klaus Ammann

University of Bern, Botanical Garden, Altenbergrain 21, CH-3013 Bern, Switzerland

Biogeographical assay provides knowledge on differences in natural gene flow in different regions, which is essential information for an ecological risk assessment for field releases of transgenic crops. In a given region, a biogeographical assay is performed to know whether wild relatives are present in the area or not, following the principle "species by species, region by region" [1, 2]. Data sources are combined to detect natural hybrids between crops and related wild species. As a synthesis, risk categories for natural gene flow are defined for the test region [1; see K. Ammann, this volume]. To obtain basic data for an ecological risk assessment and a long-term monitoring accompanying field releases and large-scale cultivation of GMP's, biogeographical distribution and hybridisation potential of crops and related wild species have to be studied on non-transgenic plants, ideally before first field releases have been persecuted.

Basic data sources for species distribution, wild relatives, hybridisation and habitat characteristics are: distribution maps for plant species (from regional to continental scale), the local and regional floristic literature and herbarium sheets of the region. Combination of these data gives a first overview on potential natural gene flow in the region and detects wild relatives which might show natural gene flow with crops.

For an estimation of natural gene flow, information on the hybridisation potential cannot exhaustively be found in the literature. Herbarium sheet survey and field excursions are essential to create a picture of the real situation in the test area.

Through screening of historical herbarium sheets by morphometrical character analysis one may detect natural hybrids found in the region. Large herbarium collections preserve dried and pressed plants, most of them collected since 1850. Field-botanists have an eye for unusual individuals which often reveal to be hybrids. Such rarities are often inserted in herbaria, consequently such collections represent fairly exactly in a qualitative way the status quo on natural hybrids found in a given region – and this not just for a 3-year project period, but over many decades. This is why you can interpret such data as long-term data. Normally, herbarium sheet labels contain valuable information about the locality, the habitat and the date. This allows drawing of distribution maps from present and past times and will be a valuable help for modellers. Such long-term data is very rare, nevertheless highly needed to develop and test model performance.

To enhance precision of evidence for natural gene flow, extensive field excursions should be performed and local agricultural experts and plant specialists contacted. Field excursions are planned in regions where data from the sources mentioned above indicate the potential presence of natural hybrids, or at least an overlap of the distribution area of both crop and related species. Reading detailed topographic maps is essential for an expedient screening of a region. On excursions by car, populations of crop species, wild relatives and of hybrid zones are mapped for later comparison with information from other sources. Interesting populations containing putative natural hybrids are sampled for further examination and hybrid determination by morphological character analysis and genetical analysis, e.g. by isozymes or RAPD.

Combination of the data sources mentioned above may reveal historical aspects of gene flow and hybridisation as cases of successful invasion or decline of a species. Spatial differentiation of the probability of natural gene flow between crops and related wild species may be detected. In regions with poor data sources, e.g. where no distribution maps are available, extensive field excursions, although being expensive, are important to assay the risk of natural gene flow from crops to potential wild partners.

The case study of the rediscovery of wild rape, *Brassica rapa* L. ssp. *campestris* Clapham (syn. *B. campestris*) (drawing of Swiss material in [2]), in Switzerland shows that the combination of data sources may detect deficiency in single sources [3]. Analysis of European distribution maps reveals that this species is not present in Switzerland and in Great Britain; obviously, this weed was not mapped there as it is too closely related with crops and therefore not interesting for plant specialists. In the majority of modern floristic books, it is not mentioned as well. As recent data are lacking, historic sources are important indicators where this species lived and where it might have survived. Herbarium sheets from the beginning of the century as well as old local floristic books prove that wild rape was a frequent weed in cultivated fields, especially in alpine valleys. In 1995, on field excursions in regions with historic locations of wild rape, we rediscovered the species in similar habitats as it grew decades ago: in traditional small-scale fields of potato and spring barley at about 1400 m in the Valais (Obergoms) and Graubünden (Unterengadin and Vorderrheintal). There, it is a frequent, but harmless weak weed, strictly restricted to cultivated fields. Oilseed rape (*Brassica napus*), with which it might form natural hybrids, is not cultivated in these regions as the climate is too rough. For a risk assessment of gene flow of oilseed rape, wild rape is therefore a minor problem in Switzerland [3].

The case study of purple flowering *Medicago sativa* (alfalfa) and yellow flowering, wild *M. falcata* (sickle medic) in Switzerland shows a regional differentiation in the probability of hybridisation due to two chromosome types of *M. falcata* [3–5]. All cultivated and feral alfalfa individuals analysed in Switzerland were tetraploid. On the other hand, we detected both tetraploid and diploid *M. falcata*, regionally separated. Extreme interspecific gene flow exists between the tetraploid types. Hybrids and introgressive forms of any flower colour shade (including often greenish flowers) are very frequent where both species meet. Genetic pressure from cultivated alfalfa is so strong that the wild type sickle medic has disappeared from large areas – this without genetic engineering. *M. falcata* is today an endangered species, threatened genetically by introgression and by intensive agriculture. Our field excursions showed that this decline goes on, and that sickle medic has disappeared from regions where it was mapped in the 1960s. Many large hybrid zones contain a wide range of hybrids, but no or very few individuals of pure *M. falcata*. Only in one single alpine valley in the East of Switzerland (Unterengadine), did we detect no hybrids, even where both parents were adjacent. Herbarium sheets and local floristic literature confirm that hybrids are very rare events there. Chromosome countings show that *M. falcata* is diploid in this region [4, 5]. But neither published distribution maps nor modern floristic literature mention this aspect; field excursions and the study of herbaria are therefore essential. For risk assessment, the chance of natural gene flow from alfalfa to sickle medic must thus be regionally differentiated.

Analysis of hybrid zones of *Medicago* is demonstrated with an example of a natural mixed population in Bonaduz, Graubünden, Switzerland [4, 5]. Measuring tapes were spread over the population forming a grid of X/Y coordinates to determine the position of the sampled plants.

For each m^2 of the transect, one plant was sampled. In case no plant was growing exactly on the crossing of a whole meter of coordinates X and Y, the closest one was chosen. The roots of the selected sprouts were dug out (for cultivation in the Experimental Garden of Neuchâtel; biochemical analysis [5]), the shoots were cut and immediately pressed for the herbarium collection and biometrical character analysis. Flower colour was determined on living material using an international colour system. Exact location of each individual within the transect was labelled with its x/y coordinates. Both biometrical and biochemical analysis were performed on the same individuals [4, 5].

Figure 1 shows the spatial distribution of individuals within a natural population of *Medicago sativa*, *M. falcata* and hybrids *M.* × *varia* (population M4a from Bonaduz, Graubünden, Switzerland). Principal Coordinate Analysis (PCA) aggregates similarity of morphological characters and identifies hybrid types. X-axis and Y-axis are the spatial population dimensions [m]. Each circle represents an individual plant, collected at the corresponding m^2 within the population. Where no circles appear, no *Medicago* plants were present. Circle diameter represent factor 1 of the PCA (about 30 morphological characters considered) indicates the *Medicago* type. Plants with factor 1 approx. >0.1 are feral plants of alfalfa (*M. sativa*) of the adjacent alfalfa field. Hybrids are found around zero on the z-axis, whereas pure individuals of *M. falcata* have a factor 1 approx. < –0.2.

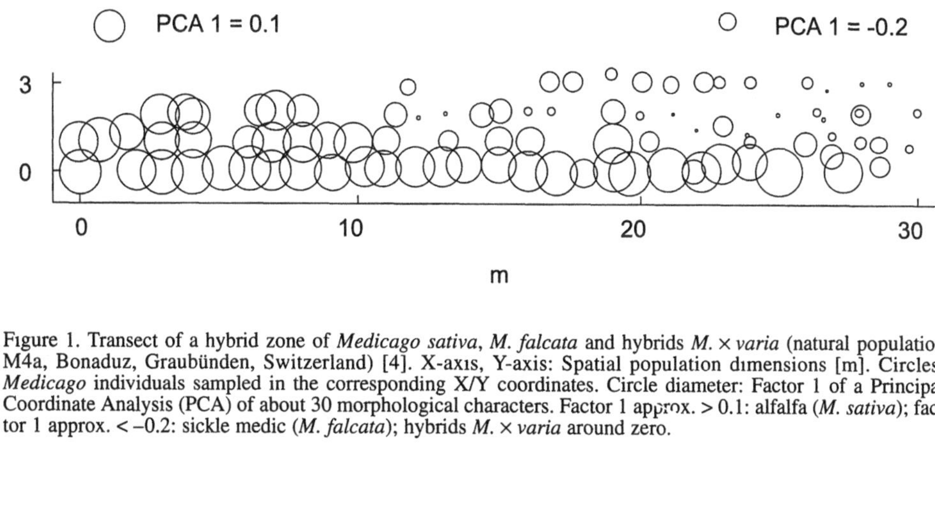

Figure 1. Transect of a hybrid zone of *Medicago sativa*, *M. falcata* and hybrids *M.* × *varia* (natural population M4a, Bonaduz, Graubünden, Switzerland) [4]. X-axis, Y-axis: Spatial population dimensions [m]. Circles: *Medicago* individuals sampled in the corresponding X/Y coordinates. Circle diameter: Factor 1 of a Principal Coordinate Analysis (PCA) of about 30 morphological characters. Factor 1 approx. > 0.1: alfalfa (*M. sativa*); factor 1 approx. < –0.2: sickle medic (*M. falcata*); hybrids *M.* × *varia* around zero.

The results of the study of natural hybrid zones show a continuum of various hybrids between *Medicago sativa* and *M. falcata*. According to the literature, "pure" parents are defined by legume characters and flower colours [6]. Population M4a of Bonaduz (total of 98 samples) contains 48% *of M. sativa*, 23.5% of *M. falcata*, 22.5% of hybrids *M.* × *varia* individuals; 6% were not clearly identifiable as they were sterile [4].

As a conclusion, biogeographical differentiation has to be considered in a risk assessment; especially for modelling and for the evaluation of long-term effects of transgenic crops. Biogeographical assay supplies spatial data, whereby combination of data sources is essential. In addition to the well-accepted "species by species" strategy, knowledge on spatial data "region by region" is essential for a planning of large scale field releases. For a monitoring of potential ecological long-term effects of transgenic crops, data should be collected before and after large field releases in order to detect potential differences in gene flow to wild species between transgenic and conventional crop varieties.

Acknowledgement
This study was funded by Swiss National Science Foundation (grant nr. 5002-035207).

References

1. Ammann K, Jacot Y, Rufener Al Mazyad P (1996) Field release of transgenic crops in Switzerland, an ecological risk assessment. *In*: E Schulte, O Käppeli (eds): *Gentechnisch veränderte krankheits- und schädlingsresistente Nutzpflanzen. Eine Option für die Landwirtschaft?* Schwerpunktprogramm Biotechnologie des Schweizerischen Nationalfonds, Bern, 101–157
2. Kjellsson G, Simonsen V, Ammann K (eds) (1997) *Methods for risk assessment of transgenic plants. II. Pollination, gene transfer and population impacts.* Birkhäuser, Basel, 308 p
3. Rufener Al Mazyad P (1999) Gene Flow between Swiss crops and related weeds. Morphological character analysis and biogeographical assay for the risk assessment of future transgenic crops. *Medicago sativa* L. (alfalfa) and *Medicago falcata* L. (sickle medic); *Brassica napus* L. (oilseed rape) and *Brassica rapa* L. ssp. *campestris* Clapham (wild rape). PhD thesis, University of Neuchâtel
4. Savova D, Rufener Al Mazyad P, Felber F (1996) Cytogeography of *Medicago falcata* L. *and M. sativa* L. in Switzerland. *Botan Helv* 106: 197–207
5. Savova Bianchi D (1996) Evaluation of gene flow between crops and related weeds: Risk assessment for releasing transgenic barley (*Hordeum vulgare* L.) and alfalfa (*Medicago sativa* L.) in Switzerland. PhD thesis, University of Neuchâtel
6. Gunn CR, Skrdla WH, Spencer HC (1978) Classification of *Medicago sativa* L. using legume characters and flower colors. Agricultural Research Service United States Department of Agriculture, Washington D.C., Technical Bulletin No. 1574, 84 p

Methods for Risk Assessment of Transgenic Plants
III. Ecological risks and propspects of transgenic plants
K. Ammann, Y. Jacot, V. Simonsen and G. Kjellsson (eds)

Gene flow between selected swiss crops and related weeds: risk assessment for the field releases of GMO's in Switzerland

Yolande Jacot and Klaus Ammann

Botanical Garden, Altenbergrain 21, CH-3013 Berne, Switzerland

Introduction

With the development of recombinant DNA techniques, plant breeders now have access to an astounding number of useful genes that can be inserted into a given plant's genome. Virtually, all commercially important plants are affected by this type of improvement, and the number of transgenic cultivars that are field-tested each year is increasing exponentially. Some crops are now commercialised and many more crop varieties are nearly ready for commercial release.

Hundreds of small-scale field tests have been carried out in order to assess the performance of transgenic cultivars under different field conditions. Results from these small-scale trials are sometimes cited as evidence that transgenic plants pose no ecological risks at any scale of cultivation, but this conclusion can be unwarranted for several reasons. First, they are usually conducted so that gene escape is unlikely. Second, the scale at which the tests are conducted is so small and short that undesirable effects on non-target organisms such as beneficial insects are extremely unlikely. Furthermore, the possibility that microbes, insects, and weeds will quickly evolve resistance to plant-produced antibiotics, toxins, and herbicides cannot be addressed in these tests due to the short duration and limited acreages involved. Finally, the fact generally reported that nothing happened in the field trial is not useful in evaluating the ecological risks unless these questions are the focus of carefully designed, long-term experiments.

One of the perceived risks of commercialisation of genetically engineered plant is that transgenes for fitness-related traits could be transferred to wild or weedy populations of these taxa and their close relatives, causing free-living taxa to become more serious weeds.

The purpose of the work presented here is to (1) Evaluate the probability of gene transfer from crop to wild relatives and (2) classify some important Swiss crops into risk categories.

For this purpose, (1) a bibliographic databank on gene flow between crops and their wild relatives has been built, (2) a convenient classification of gene dispersal probability from transgenic crop to the wild flora was adapted for Switzerland after an idea of De Vries et al. [1].

Bibliographic databank

This databank comprises the most relevant articles relating to this subject with an emphasis on the 30 studied plants.

The literature has been searched in large databases like BIOSYS, AGRICOLA, Current Contents, etc. and put together in an Access database. Each record contain the full reference of the article as well as species names and general keywords.

The data available in this bibliographical databank have been used to make an evaluation on the probability of gene flow for 30 important Swiss crops.

Dispersal codes

As a result of discussions in the symposium at Louverain, we propose *gene flow codes* after the idea of some Dutch authors [1, 2]. We are giving here an adapted version in order to spur discussion on a European level. We think it is desirable to establish a European classification system as proposed by Frietema De Vries [2], where some of our proposals have been adopted. It is not possible to arbitrate the crops and their wild relatives on one and the same level all over Europe: Classification work has to be done on a *regional* scale taking into account local environmental conditions, species and transgenes. This regional scale has been proposed by Frietema De Vries [2], following the well-known subdivisions of Meusel.

Here we deal only with the three first codes, but we feel strongly the necessity of a fourth code for the future: We need to assess also the *risk of the inserted transgene itself.* For this *Dg code* we need experimental approaches on all levels from a strict containment over small scale field releases to the large scale releases over long periods. For the time being there remains only the possibility of a rough estimate of how transgenes will have side-effects in the long run, and some comments are built in provisionally in code Dp (vertical gene flow). The authors are well aware of the pragmatic view they take, which is blurring the logic of the three codes already defined.

These codes are presented here in order to open debate on feasibility and organisation of such codes for future risk assessment: The codes can serve as a first rough estimate, before going into more detail for a risk assessment based on field monitoring and experimental approach, where judged necessary.

The three codes are: (1) Dp: Hybridisation and pollen dispersal index; (2) Dd: Diaspore dispersal index and (3) Df: Distribution frequency index (at present time).

Each index is subdivided in five levels with a sixth one for data too scanty or lacking at all (Categories 0 (lowest risk) to 5 (highest risk) and U (unknown)). In this last case no evaluation is possible. They are summarised below.

Classification of the codes for the dispersal of diaspores (Dd)

Category Dd 0

No chance for dispersal of diaspores to the wild: Seeds are sterile or otherwise deficient, they have lost reproductive function. No ecological effects are expected from fruiting of the cultivated plants.

Category Dd 1

Dd to the wild occurs only occasionally and under very favourable conditions, plants usually survive only for one season (advena), they are not adapted for survival in our climate. No ecological effects are to be expected regarding the Swiss ecosystem.

Category Dd 2

Chance for dispersal of diaspores to the wild is small, but under favourable and exceptional conditions possible. Further research on population dynamics seems necessary. For risk assessment the standing of the plants in the Swiss ecosystem can be of importance.

Category Dd 3

Chance for dispersal of diaspores (by spontaneous vegetative reproduction) is real; fruiting of the cultivated plant is essentially undesirable and will normally be suppressed by various methods. Further research on population dynamics is necessary. For risk assessment the standing of the plants in the Swiss ecosystem can be of importance.

Category Dd 4

Chance for dispersal of diaspores to the wild is real. Fruiting of the cultivated plant occurs normally during cultivation. Ecological effects can be expected from fruiting of the cultivated plant. For risk assessment the standing of the plants in the Swiss ecosystem will be of importance.

Category Dd 5

Dispersal of diaspores to the wild will be the rule. Fruiting occurs very frequently and is also extremely abundant. Ecological effects can be expected from fruiting of cultivated plant. For risk assessment the standing of the plants in the Swiss ecosystem will be of importance.

Category Dd U

Data too scanty or completely lacking, no evaluation possible.

Classification of the codes for Df (frequency of distribution)

Category Df 0

No plants of this species or of a wild relative, no feral populations found in nature; no ecological effects are expected from the introduction of the cultivated transgenic plant.

Category Df 1

Plants of this species or of wild relatives are extremely rare in the wild and have their stable place in the Swiss ecosystem in specific associations. No feral populations are found in

Switzerland. Chances for hybridising with the wild or feral populations are negligible. Locations to grow transgenic plants should be appropriately chosen in order to avoid hybridisation and any ecological effect.

Category Df 2

Plants of this species or of wild relatives are rare, but occur sporadically, distribution difficult to predict and essentially uncontrollable. Feral populations may exist in certain regions. Chances for hybridising with wild populations are scanty but unpredictable. Ecological effects from the introduction of the cultivated plant may be expected, but in most cases on a local scale only. Locations to grow transgenic plants should be appropriately chosen in order to avoid hybridisation and any ecological effect.

Category Df 3

Plants of this species or of wild relatives are not very common in the wild and have their stable place in Swiss ecosystem. Feral populations are known from Switzerland, but not frequent. Chances of hybridising with the wild populations exist but are small. Some ecological effect from the introduction of the cultivated plant may be expected under unfavourable conditions when cultivated plants and wild relatives are not sufficiently separated. Locations to grow transgenic plants should be carefully chosen in order to avoid hybridisation and any ecological effect.

Category Df 4

Plants of this species and their wild relatives are not frequent but well distributed over the whole Swiss plateau, chances for hybridising with wild populations are considerable, but under very favourable conditions it can still be safely prevented. Feral populations are known and distributed over an important part of Switzerland. Locations to grow transgenic plants should be carefully chosen in order to avoid hybridisation and any ecological effect. Detailed biogeographical studies are necessary to reach this goal.

Category Df 5

Plants of this species and their wild relatives are common and well distributed over all of Swiss plateau, chances for hybridising with wild populations must be expected and cannot be prevented in field experiments. Feral populations are frequent and distributed over the whole Switzerland. In exceptional cases locations to grow transgenic plants can still be carefully chosen in order to avoid hybridisation and any ecological effect. Detailed biogeographical studies are necessary to reach this goal.

Category Df U

Data too scanty or lacking at all, no evaluation possible.

Classification of the codes of dispersal of pollen (Dp)

Dispersal of pollen and hybridisation potential, including a differentiation of possible negative ecological effects of the inserted gene itself.

Category Dp 0

No chance for hybridisation because there are no wild relatives growing in Switzerland. No ecological effects when the cultivated plants come into flower.

Monitored field releases possible, no containment experiments and no field experiments necessary.

Category Dp 1

No chance for hybridisation with wild relatives because it is experimentally proven that wild species of the same genus in Switzerland are not compatible with the cultivated plant: (artificial pollination methods and/or embryo rescue are necessary to produce hybrids).

No ecological effects when cultivated plants come into flower. Monitored field releases possible without containment. However, experiments should be carried out, to test that there are no negative effects on the host /predator system in case of transgenes introducing new resistance and/or competition effects.

Category Dp 2

No chance for hybridisation with wild relatives because there is no record of spontaneously formed hybrids of the cultivated plant with wild species of the same genus in Switzerland.

However, hybridisation is possible under experimental conditions and progeny is fertile without any artificial help. Chances of gene flow by hybridisation are small due to various outcrossing barriers (competition, biogeographical or ecological incompatibility), but under special local or artificial conditions in agricultural systems still to be considered as possible rare events.

a) In certain species groups there is a small chance of getting new transgenic hybrids, but no invasions are to be expected.
b) In other species groups there is a small chance of getting new transgenic weeds which tend to be aggressive and will possibly cause invasions under unfavourable conditions.

Category Dp 3

Natural hybridisation occurs only occasionally, backcrosses have not been observed up to now. Local situations have to be studied carefully in risk assessment of field experiments. Species to species, region by region and step by step approach required.

a) In certain species groups and under unfavourable circumstances gene flow by pollen transfer will occur, but new transgenic hybrids do not tend to be invasive.
b) In other species groups and under unfavourable circumstances gene flow by pollen transfer can influence ecosystems negatively: Local invasions of new transgenic weeds will occur.

Category Dp 4

Chance for natural hybridisation is medium; backcrosses have been observed, successful outcrossing occurs fairly often. Natural fertile hybrids are sometimes observed, small hybrid populations can be detected in nature. Species to species, region by region and step by step approach required.

a) Transgenic hybrids will have no ecological effects on the flora of the Switzerland, since the new hybrid is only capable to invade small ecological niches, and therefore does not demonstrate any disturbing invasiveness, since the inserted gene itself did not show negative ecological effects in long-term monitoring experiments. Experiments should also be carried through proving that there are no negative effects on the host/predator system.
b) Transgenic hybrids will have ecological effects on the flora of the Switzerland, since the new weed is capable to invade ecological niches, and therefore is potentially demonstrating invasiveness. There may also be negative effects (e.g. more competitive, more allelopathic) caused by the inserted gene itself.

Category Dp 5

Chance for natural hybridisation is high; vertical gene flow occurs often, hybrids are fertile and backcross frequently. Hybrid populations are often found in nature. Species to species, region by region and step by step approach required.

a) Transgenic weeds will have no ecological effects on the flora of Switzerland, nevertheless the new weed is capable to invade important ecological niches and it will act as a new weed (which should by all means be avoided!), but the inserted gene itself does not show negative ecological effects.
b) Transgenic weeds will have negative ecological effects on the flora of Switzerland since it is capable of invading many ecological niches as a major new weed and/or since the inserted gene itself may have characters demonstrating negative ecological effects.

Category Dp U

Data too scanty or lacking at all, no evaluation possible.

Dg Codes

Transgene may affect persistence by altering traits such as seed dormancy, seed germination, tolerance to biotic (living organisms), or abiotic (non-living) stresses, or competitiveness of vegetative plant parts. These changes might be manifested in the transgenic plant's appearance as a weed in the next growing season.

Any modification that enhances population growth, such as increased reproductive capacity or survival, theoretically increases its invasiveness. Transgenes like herbicide-, insect-, salt- and disease-tolerance may provide an advantage to the crop and increase its weediness. Transgenes judged less likely to affect weediness traits are the ones that delay fruit ripening, produce pharmaceutical chemicals, or alter oil, carbohydrate, or protein composition of the seeds. However, one could reasonably ask whether the changes in seed composition affect seed-related weediness traits, such as dormancy and germination capacity [3].

The long-term persistence of fitness-related genes depends on the balance between the cost of expressing the phenotype and the strength of selection favouring the trait.

Classification by combination of the three codes

After an evaluation of the three single factors (see above, dispersal codes), the combination of these codes enables us to estimate impact of a transgenic species on the environment. Six categories of risk probability have been developed:

(1) No effect
No related species or no compatible related species of the crop are known in Switzerland. Field releases of species belonging to this category are possible without any containment or short-term monitoring. Certain transgenes have to be tested in medium-term field experiments regarding their secondary effects on ecosystems: Sustainable resistance must be achieved. To reach this goal long-term monitoring is required.

(2) Minimal effects
No records of spontaneous hybridisation between the crop and the wild relatives are known in Switzerland. Field releases are possible after a thorough clarification of the biogeographical situation. Short-term monitoring in confinements should be done prior to large scale field releases. Certain transgenes have to be tested in medium-term field experiments regarding their secondary effects on ecosystems (pest and insect resistance genes).

(3) Low but local effects
Gene flow occurs towards wild or feral species existing also outside agricultural environment and control. Release experiments should first be done in confinements and afterwards in small scale releases closely monitored. This statement is restricted to transgenes not causing enhanced competitiveness outside agricultural environment, such as herbicide tolerance. Any other transgenes should be carefully tested in confinements.

(4) Substantial but local effects
Gene flow is high and substantial, but still locally controllable. Field releases could be done within strict confinements. A case-by-case analysis including the potential effects of the transgene is required before any field releases are done. Long-term monitoring of field releases under strict biological or geographical confinement conditions is necessary in order to study competitiveness of the transgenic crop. Risky transgenes have to be avoided.

(5) Substantial and wide-spread effects
Gene flow is high, substantial, and widespread and will not be controllable by any means. No field releases of species belonging to this fifth category are possible. Medium-term monitoring under strict confinement conditions is necessary in order to find out about competitiveness of the transgenic varieties. Experiments with less risky crop varieties (e.g. with male sterility) having the same favourable effect desired.
(6) Unknown (one of the three codes is unknown)
More studies are needed before any field releases are done.

Conclusion

Because we do not yet know enough about the potential for natural hybridisation, we must regard the information presented in Table 1 as tentative, pending further research. The number of low risk crop species is probably substantial, but until further studies are conducted on a case by case and region by region basis, conclusions may be premature. Some species such as tomato, potato or maize do not appear to interbreed with wild species in Switzerland, but grassy crop species do interbreed freely with their wild relatives.

The hundreds of small-scale field tests in order to evaluate the performance of genetically engineered crop varieties are up to now not designed to investigate the ecological risks of widespread commercialisation (1994 International Symposium on the Biosafety Results of Field Tests of Genetically Modified Plants and Microorganisms in Monterey, CA, USA).

In order to achieve sustainability in cultivating transgenic crops, the focus should be on long-term monitoring of several years in the same field where the transgenic crop was planted. To assess invasiveness, the transgenic plant's capacity to disperse and establish in adjacent and nearby habitats should be investigated.

If genetic exchange between transgenic crops and wild relatives has weediness potential, there should nearly always be evidence of this process with non transgenic crop/weed complexes. Recent evolution of weed beets in France [4] demonstrate the novelty and effectiveness of certain fitness-related transgenes. Closer attention should be paid to possible effects on free-living wild relatives.

Rigorous studies of the sexual compatibility of crops and wild relatives are clearly needed to determine whether escaped transgenes are likely to persist in wild populations. Further research is also needed to predict how escaped transgenes are likely to affect the abundance and invasiveness of the transgenic hybrids.

Genetic exchange between crops and their wild relatives is known to have occurred in the past, but most often the focus of such studies has been on how the crop cultivar is affected by wild type genes rather than *vice versa*. Very little is known about the long-term persistence of crop genes in wild populations or the impact of fitness-related crop genes on the population dynamics of weedy relatives.

Because Switzerland's ecosystems do not fit with the political borderlines of Germany, France, Italy, Austria and Liechtenstein, it is appropriate to examine outbreeding causing new weediness on a more international scale, thus adding to the safety evaluation also in our small country.

Table 1. Risk categories for 22 important Swiss crops

		Dp code					
Df code	Dd code	0	1	2	3	4	5
0	0						
	1	Tomato					
	2	Tobacco					
	3						
	4						
	5						
1	0						
	1						
	2		beet				
	3						
	4						
	5						
2	0						
	1						
	2				Endive		
	3						
	4				Turnip		
	5						lettuce
3	0						
	1						
	2				Cabbage		
	3				Radish		
	4						
	5				Rape		
4	0	Maize					
	1			barley			
	2			wheat		carrot	
	3			rye	Chicory		
	4						
	5						
5	0						
	1	Potato					
	2						
	3		clovers				
	4						alfalfa
	5						grasses

No effect
Minimal effect
Low but local effect
Substantial but local effect
Substantial and wide-spread effect

Acknowledgements
The authors thank the Federal Office for Environment, Forests and Landscapes and the Swiss National Science Foundation (Priority Programme Biotechnology) for their support.

References

1. Vries de FT, van der Meijden R, Brandenburg WA (1992) Botanical files. A study of the real chances for spontaneous gene flow from cultivated plants to the wild flora of the Netherlands, *Gorteria* Supplement, Rijksherbarium Leiden
2. Frietema-De Vries FT (1996) *Cultivated plants and the wild flora. Effect analysis by dispersal codes.*PhD thesis, Rijksherbarium, Hortus Botanicus, Leiden, 222p
3. Rissler J, Mellon M (1993) *Perils amidst the promise. Ecological risks of transgenic crops in a global market.* Union of Concerned Scientists
4. Boudry P, Morchen M, Samitou-Laprade P, Vernet P, Van Dijk H (1993) The origin and evolution of weed beets: consequences for the breeding and release of herbicide-resistant transgenic sugar beets. *Theor Appl Genet* 87: 471–478

Methods for Risk Assessment of Transgenic Plants
III. Ecological risks and propspects of transgenic plants
K. Ammann, Y Jacot, V. Simonsen and G. Kjellsson (eds)

How do the design of monitoring and control strategies affect the chance of detecting and containing transgenic weeds?

Michelle A. Marvier, Eli Meir and Peter M. Kareiva

Department of Zoology, Box 35-1800, University of Washington, Seattle, WA 98195 USA

Monitoring is a compromise response to environmental debates

Environmentalists are concerned with the risks associated with transgenic crops, whereas many promoters of biotechnology can see only benefits [1]. As with many such debates, the solution is a compromise, which may satisfy neither side. One key to making both parties agree to such a compromise can be an effective monitoring program. From an environmentalist's perspective monitoring can serve to alert the public that there is indeed an ecological problem, in spite of our best hopes. From a biotechnologist's point of view, monitoring can substantiate that no ecological problems have been forthcoming, and that thus regulations should be relaxed. Monitoring needs to be sensitive enough that it triggers an alarm *before it is too late*. On the other hand, monitoring must not be so unrealistically sensitive to the point that it could never yield a verdict of safety. In this paper we explore the promise and limitations of monitoring for escaped transgenic plants that are on the verge of becoming serious weed problems. More specifically, the ecological risk at which our monitoring program is aimed is simply the establishment and spread of a novel plant. Although this is a simplistic "risk", it might well be a practical target for transgenic monitoring, since without spread and proliferation transgenes are unlikely to produce substantial impacts.

The nature of transgenic modifications in agriculture and the perceived risks

Many different risks have been suggested for genetically engineered crops – ranging from altered and impaired soil fertility, to the creation of resistant pests, to increased weediness of wild relatives [1-4]. We focus in this paper on the creation of new weed problems, either because the cultivar itself has an enhanced weedy potential when it escapes from cultivation as a result of some transgene, or because the transgene becomes incorporated into a weedy wild relative and confers more aggressive traits on that weed. Since many crops are being modified towards greater stress tolerance, the issue of enhanced weediness is a relatively general risk that warrants broad consideration with respect to agricultural biotechnology. Alternative "risks" may warrant different monitoring programs. But even in these different cases, we believe the development of any monitoring program would benefit from the type of exercise we illustrate in this paper. In particular, the only way of understanding how to design monitoring strategies is to simulate specific scenarios that one hopes to avoid, and then see if with sampling error and variability, monitoring actually can help avoid the undesirable scenarios.

Analyses of plant invasions: implications for monitoring transgenic crops

An extensive, albeit largely anecdotal, literature documents the establishment and spread of hundreds of invasive plant species [6, 7]. The patterns emerging from plant introductions and invasions can provide insights about what to expect when a transgenic crop either becomes invasive or hybridizes with wild relatives to produce an invasive weed [8]. One obvious feature of plant introductions is that most introduced plants do not become weeds. In fact, the vast majority of introduced plants fail to establish naturalized populations [9]. This should not, however, provide a false sense of security because those introduced plants that do flourish often cause immense ecological and economic damage [8].

Even more disconcerting is the finding that many, if not most, of the worst invasive plants were deliberately introduced to provide some benefit to humans [10]. For example, in northern Africa, almost half of 87 naturalized plant species were deliberately introduced [11]. In Australia, of 463 intentionally introduced plant species, 60 became listed as weeds, while only four species have been found to be useful without also being weedy [12]. Finally, in the United States, the U. S. Department of Agriculture (USDA) may be responsible for introducing most of the nonindigenous species currently identified as invasive, and in Florida, 90% of the 94 most problematic weed species were introduced intentionally [13]. Obviously, intentional introductions of plants are not necessarily benign, and caution is warranted whenever plant genotypes are introduced to novel environments.

It is generally impossible to predict invasion success

Complicating matters further is the lamentable reality that it is exceedingly difficult to identify *a priori* which species will cause economic and ecological problems. Lists of traits shared among weedy species [14] are so riddled with exceptions and ambiguities as to be largely useless. Indeed, a number of comparative studies have failed to identify simple, general predictors of invasion success [15–18]. More recently, it has been recognized that plant species that are successful invaders in one region are likely to be successful in others [19, 20]. However, at least initially, this trait will not be of much use in evaluating the invasion potential of transgenic plants due to their high degree of novelty. Nevertheless, once a transgenic crop is identified as invasive in one region, it would be wise to prohibit its introduction to other areas and to eradicate it from places where it already exists.

Time lags in plant invasions can be staggeringly long

A striking feature of plant invasions is that some invasive species quickly become widespread and troublesome pests, whereas others maintain innocuously small populations for long periods before becoming a problem [21–25]. *Mimosa pigra*, for example, was introduced to Australia between 1870 and 1890 and remained a minor weed until a dramatic increase in the late 1970 s when dense monospecific stands of *M. pigra* expanded over large areas [26]. Explosive population growth and range expansion of alien species may be triggered by environmental change or the appearance of favorable genetic traits in the invasive species [24, 25]. Tamarisk, for example, was planted to provide shade to settlements along a central

Australian river. It remained a useful and non-invasive plant for decades until a periodic flood dispersed tamarisk seeds to new habitats and opened up new sites for settlement [27]. Tamarisk then began to spread rapidly, forming dense stands and displacing native species. Remarkably long time lags between the time of first introduction and initial detection of weed spread have been documented in an extensive reconstruction of historical plant invasions [24]. Kowarik [24] used horticultural and floristic records dating back to 1594 to reconstruct the invasion history of 184 woody plant species introduced to Bradenburg, Germany. Due to a strong gardening tradition and a rich history of floristic research, the date of first introduction was fairly well documented for these species, almost all of which were intentionally introduced. Kowarik also searched for the first record of a plant having become naturalized (i.e. the occurrence of non-cultivated plants). Overall, the average time lag between the date of first introduction and the first record announcing that the plant had become naturalized was an astounding 147 years.

Data for invasive forbs are not as complete as those for woody species, but one of the best data sets comes from a reconstruction of the temporal and spatial spread of weeds in the northwestern United States [28]. Forcella and Harvey [28] used county and state weed records and herbarium specimens to construct a history of weed invasions into the northwest for five states and 90 different species, between 1890 and 1980. Unlike the ornamental shrubs and trees introduced to Germany, the date of first introduction is unknown for these weed species. Thus, we defined the detection lag as the time from first report of the plant until it was reported to have spread to 10 and 20 countries. For those species that had spread to at least 20 countries by 1980, half required 20 years to spread to 10 countries and half took over 50 years to reach 20 countries. Thus, long lags between introduction and detection of weed spread are found in both woody species and forbs. An important consequence of time lags in the detection of weed invasions is that past performance of an invader reveals little about its future potential for population growth and spread.

Lessons for monitoring transgenic plants

The history of plant introductions suggests that caution is warranted in the release of transgenic plants. Although most introduced species do not become naturalized and even fewer become economic or ecological pests, the majority of the most noxious weeds were in fact introduced to provide some economic or aesthetic benefit to humans. Further, we cannot predict which plant species will eventually become weeds. Certainly some modifications to plant traits are more likely to cause problems than others, but there simply is little about the target crop itself or its wild relatives that could allow robust predictions about its potential to become a pest. Finally, time lags in the detection of plant invasions make it difficult to evaluate the potential extent and effects of a transgenic weed. Some argue it is unfair that novel transgenic plants receive such attention, when traditionally bred plant varieties have not been so carefully scrutinized. We believe that history tells us all novel organisms should be looked at cautiously before being released into a new environment.

Finally, the cost and difficulty of controlling a weed population are greatly exacerbated once the weed population becomes well established [23, 29], and early investment may allow eradication that becomes impossible once a weed has spread [30]. Currently, there exists no reason to expect that the eradication of a transgenic weed will be cheap or easy. Thus, to minimize the cost of eradicating a transgenic weed, monitoring programs should strive toward the earliest possible detection and elimination of transgenic weeds.

Using a simulation model to assess monitoring strategies

One way of evaluating monitoring algorithms is to simulate an invasion process that is well-specified, and to ask how well the monitoring detects an invasion early in the process. There are three components to our simulation: (1) the actual model of transgenic weed establishment, population growth, and spread; (2) a simulation of sampling the environment and then analyzing data from those monitoring samples, and; (3) a simulation of control measures embarked on after an "alarm" is raised by the monitoring results.

We describe each of these separately below, but first it is important to discuss our key simplifying assumptions – both their rationale and implications. We do not intend the simulation to represent any particular weed invasion, but rather to indicate the general problems encountered when monitoring the landscape surrounding fields in which transgenic crops are grown. First we decided to represent the world as two distinct habitat types: suitable and hostile. In hostile habitats, the transgenic weed cannot live. In suitable habitats, the weeds can live and reproduce, but at rates that vary about some median from year-to-year and place-to-place. Secondly, we do not track gene frequencies or hybridization events. The spread of "genes" will be essentially the same as that of weed phenotypes, with the exception that increases from low frequency could be exceptionally slow if the selective advantage of the transgene were recessive. Rather, our starting point is the creation of some novel weed type that can thereafter breed true and spread. Thus, our model is literally accurate if one imagined a polyploid formed by hybridization which was then reproductively isolated, and it would also approximate well the spread of a dominant transgenic allele. For alternative genetics, the lag times between hybridization and weed outbreak would be greatly enhanced relative to the lag times recorded in our simulations, but the monitoring principles we uncover are general to invasion in a variable environment. Our model and monitoring program apply generally to any weed invasion with the special detail that the initial origin of the first weed innoculum is centered about a central source crop with exponentially declining probability as one moves away from the source crop.

Spatially explicit plant population growth and spread in a stochastic and heterogeneous environment

We developed a cellular automata model that simulated plant growth, reproduction and dispersal in a heterogeneous environment (Fig. 1). Our world consisted of 10 000 total cells in a 100 by 100 grid. We envisioned an annual transgenic weed (which could be a feral plant or a closely-related weed that had acquired some fitness enhancing gene through hybridization) that appears randomly, with exponentially declining probability as one moves away from a central point. This point could be thought of as an agricultural field or valley, around which monitoring would be focused. When weeds appear, they then start to reproduce and spread from cell-to-cell. Of course, an incipient invasion could die out as well – spread being certain only when the mean net reproductive rate is quite high, and variability in plant success is low.

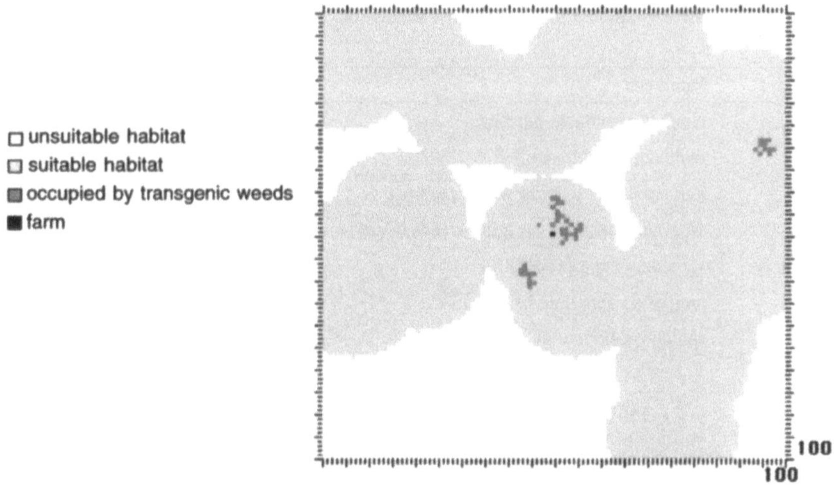

Figure 1. Snapshot of a transgenic weed simulation. Patches of suitable habitat are positioned at random, with the farm always located at the center of the grid. Transgenic weeds escape and disperse close to the farm according to an exponential distribution. The weeds cannot live in unsuitable habitat. In suitable habitat, the weeds reproduce at a variable rate and disperse, with the majority of dispersal occurring locally. There is a small chance of long distance dispersal which can lead to more peripheral weed populations such as seen in the upper right portion of the picture.

To simplify matters, the environment is thought to consist of only two types of habitat patches: hostile and potentially favorable. If seeds land in hostile patches they die. If seeds land in favorable patches, they can survive and multiply at some variable rate. The spatial deployment of favorable habitats is described by two attributes: the number and size of good habitat patches. The actual locations of favorable habitats is determined randomly at the beginning of each simulation. For the habitat parameters used here (Tab. 1), 75% of the landscape was suitable habitat, on average.

We imagine for each year the possibility of an escape or hybridization at a rate, *a*. Thus, if $a = 0.5$, there may be an escape and colonization once every other year, on average. To position the escaped plant on the landscape we assumed it would most likely fall close to the source according to the exponential curve: distance $= -\ln(x)/b$, where x is a uniform random deviate between 0 and 1, and $1/b$ is the mean distance of seed or pollen flow away from the source. We imposed the additional limitation that colonization is restricted to favorable habitat. Thus, the actual annual probability of escape depends on the random arrangement of habitat patches as well as the parameters *a* and *b*, described above.

After the plant has appeared, its "weediness" is captured in the following life history attributes: (1) the median number of offspring plants produced per "cell" in a favorable environment; (2) the fraction of offspring dispersed away from their natal site; and (3) to what extent dispersal is local (only to neighboring cells) *versus* global (spread randomly with uniform probability across the entire landscape). Each cell may contain a breeding population of plants, and the model ignores within-cell population dynamics. The median reproductive rate varies from year to year according to a lognormal distribution with specified coefficient of variation (cv). Once a yearly median has been selected, the net reproductive rate also varies from patch-to-

Table 1. Values of parameters explored in the presented simulations

Parameter type	Parameter	Values
Habitat:	number of suitable patches	25
	radius of suitable patches	14
Weed escape:	*a*, number of escaping weeds per year	0.05, 0.1, 1.0
	1/*b*, mean distance escaping weeds disperse	5
Weed life history:	*r*, median net reproduction	1.3, 1.6, 1.9
	temporal coefficient of variation	1.3
	spatial coefficient of variation	1.3
	h, proportion seeds remaining in parental cell	0.699
	n, proportion seeds dispersed to eight neighboring cells with uniform probability	0.3
	g, proportion seeds dispersed globally with uniform probability	0.001
Monitoring:	number of cells sampled 400	25, 50, 100, 200,
	1/*d*, mean distance of sampled cells from farm	10
Control:	*K*, threshold number of sampled cells occupied to initiate herbicide spray	2, 10, 20
	probability all plants in a cell die whensprayed	0.8

patch, again according to a lognormal distribution with a specified cv. To select realistic cv values we calculated the spatial and temporal cv's for the fecundities of transgenic varieties of oilseed rape reported by Crawley [31]. For the simulations reported here, we set both the spatial and temporal coefficients of variation to 130%, well within the range of cv's we calculated from Crawley's data.

The net reproductive rate per cell results from growth, survival, and seed production. Thus, when we indicate a net reproductive rate of two, that does not mean only two seeds are produced, but rather that two seeds germinate, survive, grow, and live to reproduce themselves in the next generation. We assume there is no seedbank, an assumption that makes control of the weed easier, and that slows the spread of the weed in a variable environment (since seed banks can help populations survive poor reproductive years).

In summary then, the invasion process for our hypothetical escaped transgenic entails the following factors: first, establishment of either a feral plant or a new hybrid weed is a function of the mean dispersal rate for transgenes away from central crop. A potential escapee is emitted from the source every 1/a years, on average, but whether or not this potential escapee can establish depends on whether it lands in suitable habitat. The habitat is characterized by the number and size of favorable habitat patches. Life history traits of the transgenic weed include: the median net reproductive rate in favorable habitat (r), temporal and spatial cv's, the proportion of seeds that stay "home" in the parental cell (h), the proportion of seeds that are spread randomly but with uniform probability among the surrounding eight immediate neighboring cells (n), the proportion of seeds that disperse globally ($g = 1 - h - n$) over the landscape. All of these factors could in theory influence the effectiveness and feasibility of different monitoring schemes. Each simulation was run for 100 time steps, and we ran 100 replicates for each combination of parameter values (see Tab. 1 for a complete list of parameter values explored here).

Spatially explicit monitoring simulations

Layered on top of this invasion process, we simulated a spatially-explicit monitoring scheme. The key features characterizing this monitoring scheme are the total monitoring effort (the number of cells sampled) and the spatial distribution of samples. For the simulations presented here, sampling occurs every year and we assume that sampling within a cell is completely accurate: if a weed population is present, it is detected, and any time a weed population is detected, it is truly present.

Sampling sites are randomly chosen with distance from the 'farm' described by an exponential distribution, $-\ln(x)/d$, where x is a uniform random deviate between 0 and 1, and $1/d$ is the mean distance of sampled cells from the farm, with the additional limitation that sampling is restricted to favorable habitat. Thus, the degree to which sampling sites are concentrated around the source crop depends on the mean distance of sample sites from the crop and the random arrangement of habitat patches. The arrangement of sample stations remains fixed for each run of the simulation.

Simulating control measures

Here we consider a simple scenario of how managers might decide to implement control strategies. Control is attempted any time the number of sampled cells that are occupied by weeds exceeds a threshold *K*, with *K* being a variable that is altered for different simulations. This simple trigger for control action is practical and biologically sensible: control is implemented as soon as the weed infection exceeds some threshold level. This rule does not require sophisticated statistics, and would be easy to implement.

Once the weed population exceeds the threshold level, all cells of patches where breeding weed populations were detected are "sprayed" with a herbicide before reproduction occurs. The effectiveness of this control method is given by the probability of killing all the plants within a cell, *p*. For the simulations presented here, *p* is held constant at 0.8 In other words, there is an 80% chance that the entire weed population within a cell will be wiped out by each herbicide application. We do not consider sub-lethal effects of herbicide application here.

An important consideration of developing a control strategy is the cost. The total cost of control is a function of the number of cells treated times the cost per cell, with the cost per cell given as the cost of the herbicide and application times the number of herbicide applications per year. In addition to how many cells are treated with herbicide, there is also the question of which cells are treated. Which cells are treated does not alter cost, but is something worth examining from a strategic point of view [22]. We examine both cost effectiveness and alternative control strategies in Marvier and Meir (in preparation).

Structure of our model analyses and synthesis

By combining simulated invasions with simulated monitoring programs, there are several practical questions worth exploring. First, are there some combinations of life history attributes for invasive weeds that make monitoring essentially a losing battle? Second, are there general principles of efficient monitoring design, or do these principles vary with the details of weed life history? We present our results in three stages. First, we show at what stage monitoring programs detect an invasion problem given particular sampling designs and the biological attributes of invaders. Second, we display how the severity of invasions, measured by total area "infested" by weeds, depends on life history of the weeds and the design of monitoring and control strategies. Third, we examine how the frequency of monitoring failures and the frequency of "false alarms" vary with different monitoring programs and various triggers for action. Lastly, we synthesize these results into general rules of thumb for monitoring programs.

How does the design of monitoring programs interact with biology to produce average year after initiation at which a weed is detected?

Because "detection lags" are such an inherent feature of ecological invasions, it makes sense to summarize the results of alternative monitoring programs in terms of the average delay before a weed is detected. The time to first detection is actually the sum of (1) the time until weed escape and (2) the time from first escape until first detection *via* monitoring. Thus, we explored how various weed traits and monitoring parameters influence time to escape as well as the overall time to first detection.

The weed's median rate of net reproduction has essentially no effect on the time to first detection (Fig. 2a, top). This parameter describes the rate of weed reproduction *after* it has successfully escaped and therefore has no effect on the time of first escape (Fig. 2a, middle). More surprisingly, changes in net reproduction also had little effect on the duration of the lag between escape and detection (Fig. 2a, bottom). Time to detection of a transgenic weed is slightly longer when monitoring effort is low (Fig. 2b, top). The time of first weed escape is, of course, independent of the number of cells sampled (Fig. 2b, middle), but increased monitoring effort can strongly reduce the time lag between escape and detection (Fig. 2b, bottom). The variable that most strongly affected time to detection is the rate at which weeds escape from the farm (Fig. 2c, top). This is largely caused by differences in the time to first escape (Fig. 2c, middle) rather than changes in the lag between escape and detection (Fig. 2c, bottom).

How do biological attributes and monitoring protocols influence the extent of weed infestations?

We investigated how two biological attributes of the transgenic weed, the median rate of net reproduction and the number of weeds annually escaping from the farm, affect the final size of the weed infestation. Both of these weed traits strongly affect the size of the weed infestation, measured as the number of cells occupied at the end of a simulation run (Fig. 3a). However, either reducing the threshold for herbicide spraying (Fig. 3b) or increasing monitoring effort (Fig. 3c) can greatly limit the final extent of weed infestation.

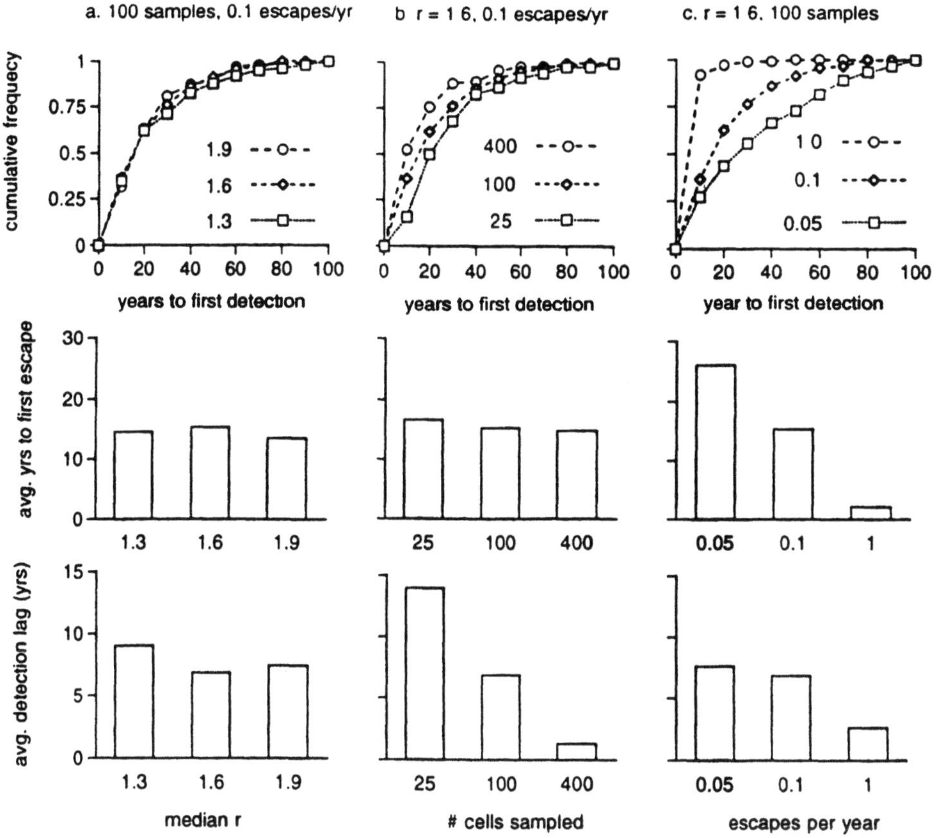

Figure 2. Time to first detection of a transgenic weed. The three columns of graphs compare how time to detection is affected by: (a) the median net reproductive rate of the weed; (b) the number of cells monitored; and (c) the number of weeds escaping from the farm each year. The top graphs are cumulative frequency distributions of time from start of a simulation to first detection of a transgenic weed; middle graphs show average time to first successful weed escape; and bottom graphs show average time lag between first successful weed escape and first detection of the weed *via* monitoring. Simulation runs where no weeds were ever detected were omitted, and the threshold for spraying was 10 cells occupied.

How do the "rules "of monitoring and control affect the likelihood of monitoring success and the frequency of false alarms?

For monitoring to be successful, the final infestation should be small. Here, we have arbitrarily defined a monitoring failure as a final infestation of greater that 100 cells occupied. Although we would optimally want the weed population to be consistently contained, examining the final time point of the simulations allows us to estimate a probability of containment. The probability of a monitoring failure depends strongly on the net reproductive rate of the weed (Fig. 4). For weeds with high reproductive rates it is very important that monitoring is intensive (Fig. 4a). *However, there is a point of diminishing returns, and certainly the economic costs of extensive monitoring may outweigh the improvements in weed containment.* Similarly, for weeds with

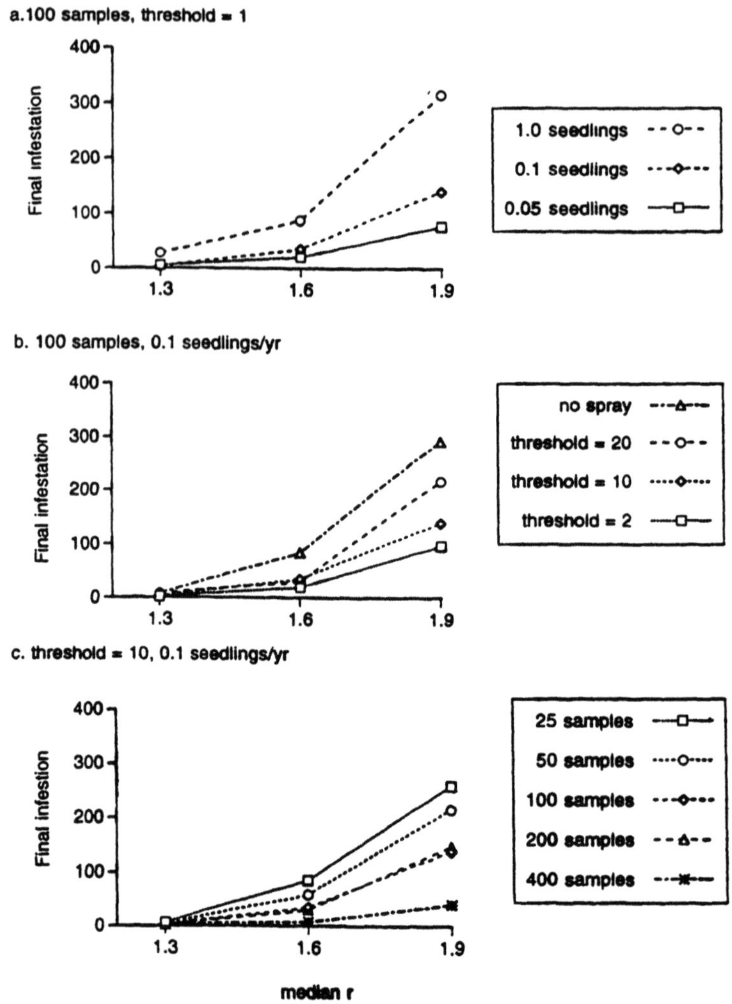

Figure 3. Spatial extent of infestation. The average number of cells occupied by weeds at time step 100 as a function of the weed's median rate of net reproduction (*r*). Lines compare how final infestation depends on: (a) the number of weeds escaping the farm each year; (b) the threshold number of occupied cells (*K*) before herbicide spray is initiated; and (c) the number of cells monitored each year.

high reproductive rates, lowering the threshold for herbicide spray can reduce the probability of monitoring failure (Fig. 4b).

Of course, increasing sampling effort and lowering the threshold for herbicide spray can also increase the frequency of "false alarms", where herbicides are sprayed even though a weed problem is unlikely. Here, we have calculated the frequency of false alarms as the difference between the number of runs in which herbicide was sprayed given particular monitoring rules and the frequency of failures in the absence of spraying (Fig. 5). The frequency of false alarms is quite insensitive to changes in the median reproductive rate of the weed (data not shown),

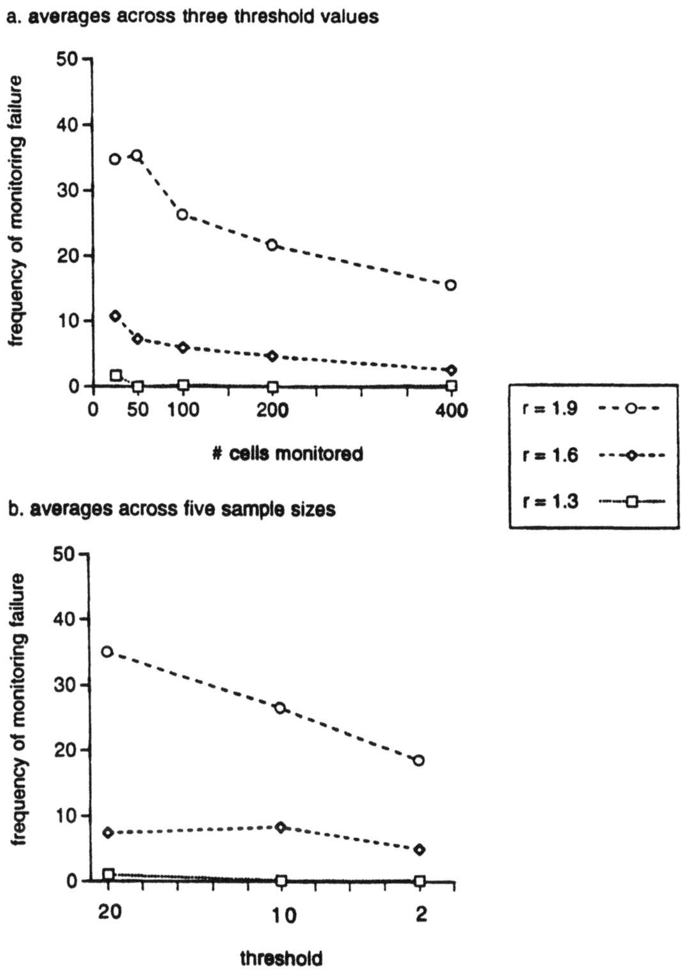

Figure 4. Frequency of monitoring failures. Failure is defined as more than 100 cells occupied at the end of a simulation run. Frequency of failure from 100 replicate runs for weeds with different median rates of net reproduction (r) as a function of: (a) the number of cells sampled each year, and (b) the threshold number of monitored cells that must be occupied by weeds to initiate herbicide spray (K).

but is strongly affected by the design of the monitoring program. In particular, low thresholds for herbicide spraying cause many false alarms regardless of the number of cells sampled. For high spray thresholds, larger sample sizes increase the number of false alarms. However, the combination of high thresholds and low sample sizes is a particularly poor strategy causing spraying to be initiated far too infrequently.

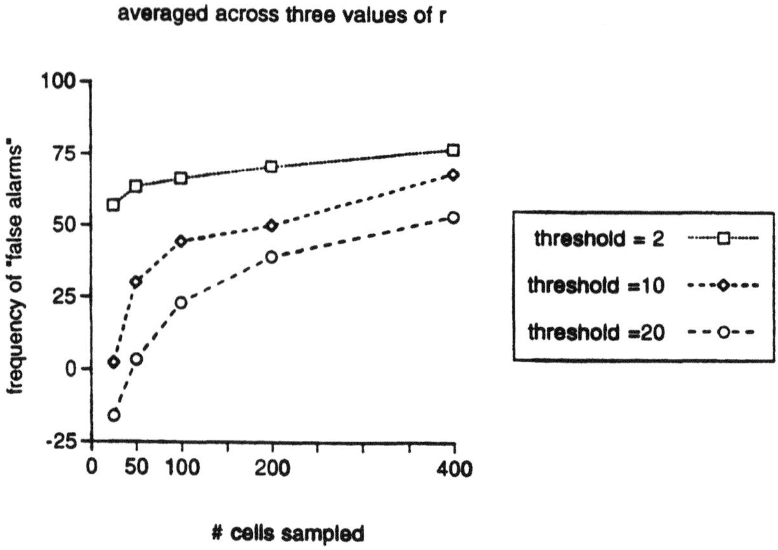

Figure 5. Frequency of false alarms. False alarms are calculated as the number of runs where spraying was initiated minus the frequency of outbreaks in the absence of spraying. Negative values indicate that spraying was initiated less often than outbreaks were occurring.

Rules of thumb for designing effective monitoring

Because detection lags can be quite long, effective monitoring will require many years of sampling. Obviously, there will be enormous pressure to discontinue monitoring efforts if no weeds have been detected after long periods of monitoring, especially if monitoring has been intensive. However, long time lags are well documented both in our simulations and in the plant invasion literature. In our simulations, a transgenic weed might not be detected for more than 90 years after the transgenic crop is first planted. In addition, sampling effort had relatively little effect on the duration of the time to first detection of a weed. The things that do affect the time to detection, such as movement of pollen and seed dispersal from the farm, are difficult to assess.

In order to guarantee detecting a problem "before it is too late", one might either emphasize extremely low triggers for an alarm to be raised (a low threshold in our simulation), or alternatively more extensive spatial coverage for the monitoring effort. Our simulations clearly indicate that over low ranges of sampling effort more spatial coverage enhances the effectiveness of monitoring as a trigger for successful control, but that these benefits of ever increasing sampling effort progressively diminish. The benefits of low triggers for an alarm to be raised are less robust – they can be dramatically influenced by the multiplication rate of the weed, and are most striking if population multiplication is quite high. This provides very practical guidance to regulatory agencies – the best guarantee of safety is widespread sampling in all cases, and a low trigger for an "alarm" if population growth rates are extremely high.

Lastly, the change in fitness due to a transgene may often appear incremental. For example, it is not clear whether a 5% or even 10% increase in fitness would seem like much of an increase.

However, we suspect that for any suite of life history traits and environmental suitability, there is some threshold for net reproduction, which once surpassed, makes monitoring and control essentially hopeless. For example, in Figure 4 it is evident that once net reproduction rises from 1.6 to 1.9, the frequency of monitoring failure becomes unacceptably high (certainly a 30 out of 100 frequency of failure would not be accepted by anyone). Where this threshold is will be difficult, if not impossible to predict. But the point is that the assurance of only small increases in fitness due to a transgene may be no assurance at all.

Can monitoring provide substantial risk reduction for transgenic crops?

Clearly monitoring, followed by control measures can offer substantial reduction in the risk of weedy invaders escaping and sweeping across our landscapes (Fig. 3, comparing "no spray" *versus* monitoring plus various spray thresholds). Just as clearly, there are circumstances in which monitoring is doomed to failure, or for which extremely broad spatial sampling is required. We are in the process of exhaustively exploring a wide range of options for monitoring and control programs in the context of different plant life histories and environmental conditions. No matter what, risk is substantially reduced by any reduction in the number of seeds (or genes) escaping from source crops. Agronomic practices that minimized this escape could combine with monitoring as a complete risk management strategy. The greatest need for study now is the characterization of temporal and spatial variability in the environment (with respect to annual plant multiplication), and of dispersal potential (see contribution by Klinger).

References

1. Snow AA, Palma PM (1997) Commercialization of transgenic plants: potential ecological risks. *BioScience* 47: 86–96
2. Anonymous (1997) EPA approves bromoxynil for transgenic cotton. *The Gene Exchange* Fall 1997
3. Metz PL, Jacobsen E, Nap JP, Pereira A, Stiekema WJ (1997) The impact on biosafety of the phosphinothricin-tolerance transgene in inter-specific *B. rapa* × *B. napus* hybrids and their successive backcrosses. *Theor Appl Genet* 95: 442–450
4. Stewart CN, Jr All JN, Raymer PL, Ramachandran S (1997) Increased fitness of transgenic insecticidal rapeseed under insect selection pressure. *Molec Ecol* 6: 773–779
5. Bigger DS, Marvier MA (1998) How different would a world without herbivory be? *Integrative Biology* 1: 1–8
6. Pysek P, Prach K, Rejmanek M, Wade M (eds) (1995) *Plant invasions: general aspects and special problems*. SPB Academic Publishing, Amsterdam
7. Stone CP, Smith CW, Tunison JT (eds) (1992) *Alien plant invasions in native ecosystems of Hawaii: management and research*. University of Hawaii, Honolulu
8. Regal PJ (1993) The true meaning of 'exotic species' as a model for genetically engineered organisms. *Experientia* 49: 225–234
9. Williamson M (1993) Invaders, weeds and the risk from genetically manipulated organisms. *Experientia* 49: 219–224
10. Williams MC (1980) Purposefully introduced plants that have become noxious or poisonous weeds. *Weed Sci* 28: 300–305
11. Le Floc'h E, Le Houerou HN, Mathez J (1990) History and patterns of plant invasion ın Northern Africa. *In*: F di Castri, AH Hansen, M Debussche (eds): *Biological invasions in Europe and the mediterranean basin*. Kluwer Academic Publ., Dordrecht, 105–133
12. Lonsdale WM (1994) Invitıng trouble: introduced pature specıes in northern Australia. *Aust J Ecol* 19: 345–354
13. Gordon DR, Thomas KP (1997) Florida's invasion by nonindigenous plants: history, screening, and regulation. *In*: D Simberloff, DC Schmitz, TC Brown (eds): *Strangers in paradise: ımpact and management of nonindigenous species in Florida*. Island Press, Washington D.C., 21–37
14. Baker HG (1965) Characterıstics and modes of origin of weeds. *In*: HG Baker, CL Stebbins (eds): *The genetics of colonizing species*. Academic Press, New York, 147–169

15. Newsome AE, Noble IR (1986) Ecological and physiological characters of invading species. *In*: RH Groves, JJ Burdon (eds): *Ecology of biological invasions*. Cambridge University Press, Cambridge, 1–17
16. Roy J (1990) In search of the characteristics of plant invaders. *In*: F di Castri, AH Hansen, M Debussche (eds): *Biological invasions in Europe and the mediterranean basin*. Kluwer Academic Publ., Dordrecht, 335–352
17. Lodge DM (1993) Biological invasions: lessons for ecology. *Trends Ecol Evol* 8: 133–137
18. Pysek P, Prach K, Smilauer P (1995) Relating invasion success to plant traits: an analysis of the Czech alien flora. *In*: P Pysek, K Prach, M Rejmanek, M Wade (eds): *Plant invasions: general aspects and special problems*. SPB Academic Publishing, Amsterdam, 39–60
19. Scott JK, Panetta FD (1993) Predicting the Australian weed status of southern African plants. *J Biogeogr* 20: 87–93
20. Reichard SH, Hamilton CW (1997) Predicting invasions of woody plants introduced into North America. *Conserv Biol* 11: 193–203
21. Hengeveld R (1989) *Dynamics of biological invasions*. Chapman and Hall, New York
22. Moody ME, Mack RN (1988) Controlling the spread of plant invasions: the importance of nascent foci. *J Appl Ecol* 25: 1009–1021
23. Hobbs RJ, Humphries SE (1995) An integrated approach to the ecology and management of plant invasions. *Conserv Biol* 9: 761–770
24. Kowarik I (1995) Time lags in biological invasions with regard to the success and failure of alien species. *In*: P Pysek, K Prach, M Rejmanek, M Wade (eds): *Plant invasions: general aspects and special problems*. SPB Academic Publishing, Amsterdam, 15–38
25. Crooks JA, Soule ME. Lag times in population explosions of invasive species: causes and implications. *In*: OT Sandlund, PJ Schei, A Viken (eds): *Norway/UN conference on alien species: proceedings*. Directorate for Nature Management and Norwegian Institute for Nature Research, Norway, 39–46
26. Braithwaite RW, Lonsdale WM, Estbergs JA (1989) Alien vegetation and native biota in tropical Australia: the impact of *Mimosa pigra*. *Biol Conserv* 48: 189–210
27. Griffin GF, Stafford Smith DM, Morton SR, Allan GE, Masters KA, Preece N (1989) Status and implications of the invasion of tamarisk (*Tamarix aphylla*) on the Finke River, Northern Territory, Australia. *J Environ Manage* 29: 297–315
28. Forcella F, Harvey SJ (1981) New and exotic weeds of Montana. II. Migration and distribution of 100 alien weeds in northwestern USA, 1881–1980. *Montana Department of Agriculture, Noxious and Exotic Weed Survey of Montana*. Montana Department of Agriculture, Helena, Montana
29. Simberloff D, Schmitz DC, Brown TC (1997) Why we should care and what we should do. *In*: D Simberloff, DC Schmitz, TC Brown (eds): *Strangers in paradise: impact and management of nonindigenous species in Florida*. Island Press, Washington D.C., 359–367
30. Simberloff D (1997) Eradication. *In*: D Simberloff, DC Schmitz, TC Brown (eds): *Strangers in paradise: impact and management of nonindigenous species in Florida*. Island Press, Washington D.C., 221–228
31. Crawley MJ, Hails RS, Rees M, Kohn S, Buxton J (1993) Ecology of transgenic oilseed rape in natural habitats. *Nature* 363: 620–623

Discussion session 4: Monitoring methods

Jarle Tufto
Why should you not release genetically modified alfalfa when the hybrid has a herbicide resistance gene or whatever. What is the problem?

Pia Rufener Al Mazyad
Alfalfa is a perennial and it seems to be very invasive in Switzerland. Anywhere you have once cultivated alfalfa you can find it several years later. I wanted to show you that the wild relative, the yellow one, the sickle medic (*Medicago falcata*) is on a decline and needs protection.

Klaus Ammann
It is just a question whether you want to have transgenes in a range of about 20 different species with a mostly southern European distribution. Yes, for some transgenes I would not care, but for others we do not know what effects they will have. I feel uneasy about high gene-flow plants to be transformed in the future. If we can avoid it, we should do so. Alfalfa is very good crop plant anyway and it does not need a lot of dramatic enhancement.

Jan Carel Zadock
You say that *M. falcata* is changing, because of introgression and I want to know when is a *M. falcata* no longer a *M. falcata* plant?

Pia Rufener Al Mazyad
Even Linné had this problem. The description of the characters is defining *M. falcata* and separating it from hybrids. My small experiments in the botanical garden crossing alfalfa pure breeding *M. falcata* showed that plants from F_1-hybrids back-crossed to the parents were really weak and had a low fitness. When pure breeding *M. falcata* is crossed with hybrids it loses fitness and therefore there is decline of *M. falcata* in the population. We also found populations with few *M. falcata* and a large range of hybrids. But the pressure on the plants of *M. falcata* described here has no connection genetic engineering.

Jan Carel Zadok
In the ethical considerations we often think about the integrity of the species. Now, is the integrity of the species of *M. falcata* damaged when the colour is breaking into pink? I doubt it, but if the multiplication possibilities are hampered, then the integrity of the species is damaged.

Simon Barber
It seems to me from the last presentation that a risk assessor in any country would be very pleased to have the sort of information available as just presented. This is very basic information, which we need. It is based on taxonomy which I think today it is a science that perhaps is not looked upon too favourably, but I see it as the absolutely key knowledge for anybody having to make sensible science-based decisions.

Gösta Kjellsson
In Scandinavia, or at least in Denmark and Sweden, we tend to have some differences between the pure *M. falcata* and the hybrids. The habitat for pure *M. falcata* is very sandy soil and hybrids tend to be more ruderal where they occur. Anyway, this is according to the traditional taxonomists and the way they see this species. Were there differences in habitat choice between the hybrids and the pure *M. falcata* in your study?

Pia Rufener Al Mazyad
The differences among habitats were that more *M. falcata* and hybrids were found together on poor soil. Nutrition could be a factor. We did not have a pure zone of *M. falcata* and mixed zones with the hybrids.

Les Lewidov
Klaus Ammann, you concluded that no release of transgenic alfalfa will be acceptable. I would like to make this explicit. P. Dale asked, what is the baseline of acceptability? If a transgenic insecticidal plant causes no more harm

than caused by chemical pesticide then perhaps it is acceptable. In this session we have to focus on the variation around the baseline. If understanding your conclusion correctly, your baseline of acceptability is literally the present form of genetic diversity in alfalfa and all its weed relatives and its particular form of diversity must not be contaminated by any gene which would not otherwise have spread in the populations.

Klaus Ammann
As long as we do not have more long-term experience with transgenes jumping around I would like to see crops with a really maximum and high gene-flow be banned from being transformed.

Phil Dale
I would like to get on to dealing with monitoring, because this is the really important thing.

Glynis Giddings
I think the one line of research which we sometimes forget about, particularly with regard to invasive species, is what an invasive species does. Does it damage perhaps by marginalizing other species in the environment and how many species are important, how important is biological diversity for maintaining community life support systems? The research has only, with a few exceptions, just really started and I think it will be important both for transgenics and for further issues in ecology.

François Pythoud
Regarding monitoring methods there are three basic questions. When do you do monitoring, how do you do monitoring and why do you do monitoring? The first presentation showed us that we can do monitoring, the methods are available. The second presentation was for me a clear case where monitoring is needed in case of high gene flow. But my question to Klaus Ammann is how can you justify a ban on transgenic crops with high gene flow with the argument of lack of knowledge on a long-term basis? This way you will never ever be able to achieve long-term experience?

Klaus Ammann
This must be a misunderstanding: My plea is for a delay of mass releases and a continuation of experimental and cautiously done field releases. My baseline of non-acceptance is, as long as we lack long-term experience, all transformation into crops with an extremely high gene flow, such as alfalfa and wild grasses. Oilseed rape does not belong to this category, but still, some precaution is necessary.

Henri Darmency
I have a comment on the first speaker today. You said that a gene is not expressed, so there is no concern. When not expressed you can have mutation, this means evolution, because the gene is not used. This needs monitoring tools. So you must come back to the information, that means to DNA, PCR and something like this.

Georg Karlaganis
It was mentioned that alfalfa is a high gene-flow plant and my question is, how high is high gene-flow? Can you suggest general parameters which make this differentiation between high, medium or low gene-flow also understandable for non-specialists.

Klaus Ammann
It is actually the Dutch group with van der Meijden who invented this classification-system of pollination gradients and seed-dispersal and frequency of a given crop in a given region, and in each of these three cases we have now defined five classes (in aberration to the Dutch system). One of the prerequisites of a 5,5,5 plant, the highest risk class, is for instance easily reproducing hybrids with viable seeds and extensive populations on hybrids, which is the case for alfalfa in Switzerland. Seed dispersal is not particularly high, but human activity is helping here. And you cannot deny that this plant is frequently cultivated here in Switzerland. The result is a crop with a high gene flow in Switzerland.

Gerard Oostermeijer
Three comments on the Dutch study, K. Ammann just mentioned.

Ruut van der Meijden and Femke Frietema de Vries started that study, where they have been looking at actual data on out-crossing on hybridization that they could find in herbarium material.

A second comment: Dutch regulators regard transgene flow *per se* not as a risk. Only in case of data showing negative effects the scheme developed by van der Meijden should come into place.

A third comment about monitoring. I think that we can talk about monitoring very easily at the moment when we still have very simple transgenic plants with very straight forward constructs in them. I think in another ten years the constructs will be much more complicated, that will be genes that are put together with much more sophisticated regulatory elements around them, and then your monitoring-questions might be different compared to these which you ask today.

Glynis Giddings
I would like to really address this. I wonder whether it is possible to search the databases and come up with a long enough unique site, which we might use as a motif when we put in transgenes, for later identifying for the DNA test, but it might be a crazy idea.

Andreas Seiter
This has been proposed and the companies had already started to design specific probes which should be put into the transgenes. However, a company is not eager to do so, because there is also the chance of misuse of these things.

Kornel Burg
It is not necessary to put in any extra. There are already unique sequence-combinations in transgenes like the promoter gene, border sequence and so on, which could be used for monitoring. We are using this sort of sequence to monitor transgene-sequences in raw and prepared food-products. If a company wants to bookmark the product, this is possible, as it is anyway unique.

Hannes Richter
Is it possible to monitor if you get soybeans in tons and there is only one percent of genetically modified, or how many kilos do you need of product to detect by a PCR-product? What is the limit which you can detect?

Thomas Nickson
I am a little confused since I thought we are actually talking about ecological risk assessment in this symposium and this is actually a question about food and feed matters. But let me give here a general answer, which leads back to what we actually should debate here:

No matter what analytical method you may use, it has to be appropriate, it has to be specific to the question and has to be valid. And by valid, I mean, that the method has to be precise, transferable and reproducible.

Session 5: Population genetics

Methods for Risk Assessment of Transgenic Plants
III. Ecological risks and propspects of transgenic plants
K. Ammann, Y. Jacot, V. Simonsen and G. Kjellsson (eds)

Transgene movement *via* gene flow: recommendations for improved biosafety assessment

Terrie Klinger[1] and Norman C. Ellstrand[2]

[1] *University of Washington, Friday Harbor Labs, 620 University Road, Friday Harbor, WA 98250, USA*
[2] *Department of Botany and Plant Sciences, University of California, Riverside, CA 92521-0124, USA*

Introduction

In a seminal publication ten years ago, editors John Hodgson and Andrew Sugden brought together papers from ecologists and biochemists regarding the "Planned Release of Genetically Engineered Organisms" [1]. At the time, experimental and commercial releases of transgenic organisms were anticipated, but few actual releases had occurred, so most of the ecological treatments were speculative, drawn from first principles, or couched in hypothetical terms. Even so, it is worth noting the prescience of some of the contributors. For example, Regal [2] argued that "Genetic engineering is improving at an astonishing pace and it is likely that biotechnologists will be able to produce a class of organisms that is viable in nature and does not consist of crippled freaks." Indeed, at this time, barely 10 years after that statement was written, it is difficult for some to remember that not long ago the products of biotechnology were expected by many to be so crippled as to be nonviable in nature.

Genetic engineering for crop improvement has advanced quickly in the past decade. Agronomic field tests of genetically engineered organisms, and particularly genetically modified plants, have increased exponentially over this same period. In the United States alone, more than 2000 field trials had been conducted as of 1996 [3]. And although the number of ecologically-based trials of various transgenic and nontransgenic entities also increased over this period, there has emerged no consensus regarding the risk of ecological detriment posed by the commercial release of transgenic products.

Our purpose here is to (1) review the evolution of ecologically-based field trials of gene flow by pollen as they relate to risk assessment; (2) comment on the efficacy of such trials; and (3) suggest improvements to existing methodologies. We restrict ourselves to consideration of the hazards posed by gene flow by pollen, and do not discuss other means of transgene escape (e.g. through direct escape and establishment of transgenic cultivars), although we realize that the ecological effects of transgene escape by various means will not be wholly separable. It is important to point out that our review of the topic is not exhaustive, but we do intend that it be representative. Throughout, we have tried to distinguish between relevant processes (e.g. hybridization, persistence, spread) and factors that will influence or modify such processes (e.g. multiple source populations, repeated introductions, population size, etc.). Variability in both of these will contribute to uncertainty in the estimation of risk.

Hybridization

Spontaneous hybridization is known to occur between a variety of traditional crops and their weedy or wild relatives (e.g. [4–13]). From what is now known of the inheritance of many engineered genes as single-locus Mendelian traits [14], it must be expected that similar hybridizations between engineered crops and their wild relatives will occur in many crop-weed systems. Indeed, spontaneous hybridization between genetically engineered crops and their weedy relatives with consequent production of viable transgenic hybrids has already been documented between *Brassica* species in diverse settings on two continents [15–18].

Given the existing evidence, new gene flow studies should seek to demonstrate the potential for hybridization in those crop-weed systems for which no hybridization data exist. Gene flow will not present a significant ecological hazard in all crop-weed systems, and it is therefore important to know which systems are inherently hazardous. Further, and more importantly, new gene flow studies should seek to determine the rate of transgene introduction to wild or weedy populations. The rate of gene flow is important because it will influence both the persistence and spread of transgenes in the wild, and these are factors of serious ecological concern.

Rates of hybridization

Although many studies have demonstrated the occurrence of hybridization, few have estimated actual rates of hybridization. Kareiva et al. [19] have pointed out that measures of hybrid frequency at distance are insufficient indicators of the probability of distance-dependent gene flow; instead, the proportion of hybrid seeds as a function of distance from the crop provides a more informative measure.

In a recent study, Timmons et al. [20] measured airborne pollen densities at distances of up to 2.5 km from fields of oilseed rape over a 3-year period. The distance-dependent proportion of transgenic hybrids in wild populations was estimated from pollen densities, and these estimates were supported with progeny analysis of wild plants. The characteristic dispersal distance for transgenic pollen in this setting was calculated to be 128–172 m, far greater than previous reports for this (and other) species, suggesting that significant amounts of pollen can travel comparatively long distances.

Multiple source populations

The dynamics of gene flow to wild or weedy populations from multiple source populations has rarely been accounted for in field trials or models (but see [20]). Indeed, Kareiva et al. [19] noted with reference to their model that the presence of multiple pollen sources would make analysis difficult. The presence of multiple transgenic pollen sources is, however, highly likely in commercial settings where a single crop comprises the dominant agricultural product within a region. In such regions, multiple plots of transgenic cultivars will almost certainly be grown in adjoining or adjacent fields, all within pollination distance of the same wild or weedy relatives. We expect that where multiple source populations exist the rate of introduction of transgenes to weedy or wild populations will increase. The rate of introduction will be sensitive to a number of variables, including the number, size, and distance of source populations. Clearly, the configuration of source populations relative to recipient populations constitutes a critical variable that will change with each application and over time. Therefore, levels of predictability when multiple-source populations are present will be low.

Of additional concern in this context is that, among multiple source populations, each could bear different genetic constructs. Thus, the potential for simultaneous escape of more than one transgene could make management and eradication efforts substantially more difficult. Further, where multiple transgenic source populations are present, the potential for crop-crop gene flow is substantial. Crop contamination has ramifications for environmental safety in that crop-crop matings could produce, in a single generation, offspring that carry multiple transgenes of unintended combined effect that are capable of hybridization with wild plants, thereby serving as stepping stones for gene escape.

Repeated introductions

The potential effects of repeated introductions of transgenic products have received little attention. Repeated introductions are likely to increase the probability of gene escape for at least two reasons. First, continuous gene flow into recipient populations could overcome the effects of selection against the transgenic hybrid phenotype. Second, the extensive literature on introduced species indicates that most introductions fail to result in the establishment of new populations [21]. However, each new introduction of a species (or transgene) constitutes an independent trial, providing a new opportunity for establishment. Given spatial and temporal variability, repeated introductions will increase the probability of unintended gene flow. Thus, in agricultural settings where the same crops are regularly replanted in the same fields, the potential for successful introduction of transgenes to wild populations will increase with the number of field releases.

Effects of donor and recipient population size

The effects of population size on the movement of transgenes from crops to wild plants are poorly known. To date, most field trials designed to estimate gene flow from crops to wild or weedy relatives have utilized variously-sized donor populations and small, often isolated or patchy recipient populations. The use of small recipient populations is presumably based on the expectation that wild or weedy plants growing within mating distance of an agricultural crop will comprise volunteers or accidental occurrences that are few in number. However, not all potential applications are adequately described by the use of large donor populations and small recipient populations. For example, both donor and recipient populations may be large, as is likely for some applications of genetically modified silvicultural species (e.g. spruce, aspen, Douglas fir) in regions where wild or unmodified congeners are abundant. Alternatively, donor populations may be small in comparison with their recipient populations (e.g. small-scale or subsistence farming of a crop within its center of diversity), or donor and recipient populations may both be small (e.g. small-scale farming in areas where congeners are scarce). Each configuration (large-to-large, large-to-small, small-to-large, and small-to-small) will have consequences for the transmission and persistence of engineered genes in the wild. Explicit tests of the effect of population size therefore are necessary.

Because little empirical evidence exists concerning the problem of transmission of engineered genes relative to population size, we look to the broader literature ([22, 23] and references therein) for general predictions. Clearly, the amount of pollen released constitutes a first critical variable. In this context, the size of the pollen population (i.e. the total number of pollen grains released) is more important than the number of plants in the donor population. Although the total amount of pollen released will generally increase with the size of the donor plant population, pollen abundance will be modified by mating system, pollen vector, timing of harvest relative to flowering phenology, and selected or engineered traits that alter pollen production. Large pollen populations will increase the likelihood of successful pollination of wild or weedy plants for the obvious reason that mate encounter is a function of density. Large pollen populations therefore present a relatively higher level of risk than do small populations.

More difficult to predict are the effects of recipient population size. Gene flow will generally increase as the recipient population decreases in size, because in small populations there are fewer potential recipients for a fixed amount of crop pollen, and because pollinators will effect fewer intrapopulation matings within small populations [23]. Therefore, the proportion of seeds sired by crop pollen will tend to increase in small recipient populations, though the number of hybrid progeny will be limited by the number of ovules available for fertilization. Conversely, gene flow will generally decrease in large recipient populations because a smaller proportion of seeds will be sired by crop pollen, though this could in fact constitute a large number of seeds because of the large number of ovules available for fertilization. Thus, high rates of gene flow into small recipient populations could result in fewer hybrid progeny than comparatively lower rates of gene flow into large recipient populations.

New theoretical work indicates that the probability of fixation of beneficial mutations will depend on changes that occur in population size [24]. These results might be extended to the case of agricultural transgenes if one equates new mutations with the introduction of transgenes to wild populations *via* hybridization, suggesting that the probability of fixation of beneficial transgenes introduced into wild populations will be sensitive to changes in the size of the recipient population. Further, the probability of fixation will depend on the number of copies of the transgene introduced into the recipient population, with more initial introductions (i.e. greater rates of gene flow) leading to a higher probability of fixation.

Effects of population size: Specific examples

Ellstrand et al. [25] estimated gene flow between several small populations separated by 255–400 meters and larger populations at distances of more than 650 meters. They found that the size of the donor population had a greater effect on gene flow than did interpopulation distance alone. Manasse [26] tested gene flow from a small donor population (n = 16) into small recipient populations of two sizes (n = 4 and n = 16). She found no effect of population size on hybridization rate. Conversely, Klinger et al. [12], using a larger donor population and recipient populations of one, two, and nine individuals found a significant effect of recipient population size on hybridization. At the crop margin, populations of one or two individuals produced more hybrid progeny than populations of nine at the same distance. This effect was reversed at a distance of 400 meters, where populations of nine produced more hybrids than populations of one or two. This result could reflect the combined effects of the number and position of available mates and the ability to attract pollinators. In any case, it should be noted that even a population of nine constitutes a relatively small recipient population, and that wild or weedy populations in commercial settings will often be much larger.

Goodell et al. [27] measured gene flow as a function of recipient population size, distance from multiple pollen sources, and cross-compatibility among mates. They found that populations of two experienced greater gene flow than populations of five or more, that gene flow into populations of two was negatively associated with distance to larger populations, that gene flow was significantly affected by compatibility between mates, and that significant interactions occurred between factors. These results imply that distant-dependent gene flow will be sensitive to a number of variables, which could render predictions of gene flow in natural settings difficult or unreliable.

Effects of "scaling up"

The effects of population size on gene flow must be considered in extrapolation of experimental results to commercial applications. As noted earlier, most trials designed to estimate gene flow have utilized small recipient populations and experimental donor populations that are smaller than their fully-commerical counterparts (but see [20]). The most significant effect of increasing the size of the donor population will be an associated increase in the size of the pollen population. Similarly, increasing the size of the recipient population(s) will increase the number of ovules available for fertilization. Therefore, gene flow could be expected to increase as the scale of release increases. Significantly, Timmons et al. [20] found that the characteristic dispersal distance for oilseed rape pollen increased from a previously-reported value of 2.3–8.8 m to 128–172 m when commercial-sized fields and more intensive monitoring efforts were employed. Clearly, the dynamics of crop-weed gene flow will best be described by the use of commercial-sized plots in relevant agricultural settings, and new studies should take this into account.

Sources of variability and problems in estimation of gene flow

Sources of variability in gene flow by pollen include species and cultivar composition and genotype, size and geometry of donor and recipient populations, specific habitat or environment, plant density, season, year, pollen vector and pollinator species composition, pollinator behavior, flowering phenology, and mating system. Variability introduced by these factors will limit the utility of single trials in generalizing to multiple applications, and could in some cases lead to underestimates of risk.

Further, the occurrence of low-frequency, large-magnitude pollination events could substantially increase the difficulty of assessing risk associated with the release of transgenic crops. The occurrence of exceptional pollination events has now been demonstrated in several studies, though their importance often has not been fully appreciated. For example, Klinger et al. [11] reported high variance in the frequency of the crop marker among pollen recipients 400 m distant from a crop of non-transgenic *Raphanus sativus*. Subsequent review of the original data showed that within one plot of nine plants, a single plant accounted for 37% of all the crop-sired pollen at that distance, with 39% of its progeny testing positive for the crop marker. In comparison, three of the plants in the same plot produced no crop-sired seeds, and five others produced between 3% and 19% crop-sired seeds. A second, separate plot of five recipients at the same distance from the crop produced no crop-sired seeds at all. Differences between plots at this distance were likely related to the directional (i.e. non-random) movement of pollinators. Differences between plants within a plot were probably attributable to differences in pollinator attraction (e.g. flower color) or interplant compatibility. Results of a second experiment in the same field the following year showed another large-magnitude pollination event, but in a different plot and position [12]. Within a plot of nine recipient plants 200 m distant from the crop, four plants showed no evidence of weed-crop hybridization, three others showed hybridization rates of 2.8–5.5%, and a single plant exhibited a hybridization rate of 26%. Hokanson et al. [28] reported a similar finding for *Cucumis sativus*, in which a single plot of recipient plants 50 m from a donor plot showed a donor-recipient hybridization rate of 38%, in contrast to seven other recipient plots at the same distance that showed no evidence of hybridization. Other examples of unusual pollination events at distance were reported for crop sorghum x johnsongrass hybrids [6], sunflowers [4, 29], millet [30], and engineered cotton [31]. Species- and cultivar-specific variability in hybridization rates have also been reported. For example, Langevin et al. [13] found that hybridization rates between five rice cultivars ranged from 1% to 7.36%, while a sixth cultivar showed a hybridization rate of 52.18%.

High variability in gene flow among replicate plots at a single distance and between individual plants within a plot suggests that the use of average rates of gene flow could result in negative bias in the estimation of risk. That is, low-frequency, large-magnitude, or exceptional pollination events could introduce genes into wild populations at rates far exceeding those predicted from measures of average gene flow, creating isolated "hot-spots" of transgenic hybrids. These hot-spots could contribute disproportionately to the spread of transgenes, because the frequency of the transgene will be disproportionately greater within them. The occurrence of exceptional pollination events therefore must be considered in the selection and application of models used to estimate the relationship between gene escape and distance from a source crop. As noted by Kareiva et al. [19], selection of appropriate models for risk assessment will often entail a compromise between underestimating short-distance gene flow and overestimating long-distance gene flow. From our perspective, model selection should be biased towards those that are best able to accommodate long-distance, high frequency, or "worst-case" gene flow. Risk-averse strategies therefore should seek to incorporate or otherwise weight model-fitting protocols to exceptional events. An important point here is that, given the fairly low sample sizes and sparse coverage of most field trials (especially at distance), the ability to detect exceptional events typically is low.

Thus, while the average relationship between the probability of gene flow and distance from a source (based on tens or hundreds of data sets) can be described by a monotonically declining function (e.g. a constant loss model), the results from individual trials may not be best described by such a model (e.g. [20]). Therefore, it is critically important to account for both the average probability of gene flow and independent observations of unexpectedly high rates of gene flow in the development of risk-averse strategies.

Persistence

Persistence of engineered genes in wild-type backgrounds will result from introgressive hybridization. Persistence can be estimated as the frequency of occurrence of the introduced transgene(s) over some number of generations (e.g. [4, 17, 32, 33]), inferred from measures of hybrid fitness (e.g. [13, 34, 35]), or from other measures such as comparative seed dormancy [36]. Clearly, direct measures of persistence are more reliable than inferences, though direct measures to date have been few.

Persistence of engineered genes in wild populations is likely if (1) the engineered trait is favored by selection and (2) costs or negative pleiotropic effects associated with the engineered trait are minimal. However, even deleterious transgenes could persist in natural populations under conditions of constant gene flow, because consistent, high levels of gene flow will overcome the effects of selection. Persistence of introduced transgenes therefore can be expected in numerous, diverse crop-weed systems under various combinations of selection and gene flow.

The size of the recipient population will affect the persistence of the escaped transgene. Transgenes may be less likely to persist in small recipient populations than in larger ones because small populations are more susceptible to local extinction, especially in agricultural settings where eradication is practiced. Local extinction or eradication of wild or weedy plants will serve to remove introduced genes from the local environment. On longer time scales, one must consider the effects of genetic drift on the persistence of transgenes. Genetic drift causes loss of heterozygosity and fixation of alleles. Drift could therefore act to either increase or decrease the frequency of transgenes in wild or weedy populations, with small populations being most susceptible to either fixation or loss.

Specific examples of persistence

Whitton et al. [4] followed the persistence of two non-transgenic, crop-specific RAPD markers in wild sunflowers (*Helianthus annuus*) for five generations. Spontaneous hybridization between the crop and nine wild sub-populations occurred in the first generation, after which time the crop was disposed of. Cultivar allele frequencies showed no significant decline over the following four generations, although significant effects of distance and generation were detected. The authors concluded that (1) cultivar alleles were able to escape and persist at moderate frequencies in naturally-occurring sunflower populations, and (2) neutral or favorable transgenes will have the potential to escape and persist in wild sunflower populations.

Metz et al. [32] followed the fate of a phosphinothricin-tolerant transgene in *Brassica rapa* × *B. napus* hybrids under carefully controlled glasshouse conditions. They found that the engineered construct was readily transmitted to F_1 hybrids and that the construct retained activity in hybrid progeny. The engineered trait was expressed by about 10% of the BC_3 and BC_4 generations, implying that the construct was stable across several generations. Mikkelsen et al. [17] found that spontaneous hybridization between *Brassica napus* and *B. campestris* (= *B. rapa*) engineered for glufosinate resistance produced viable transgenic progeny as early as the first-backcross generation, and that activity was retained in the second-backcross generation. Chèvre et al. [33] found that engineered resistance to the herbicide glufosinate ammonium persisted through four generations of *Brassica napus* × *Raphanus raphanistrum* hybrids, but suggested that transgene introgression within the weed genome would occur slowly and at a low probability because four generations were required for the transgenic hybrids to achieve a chromosome number and morphology close to that of the weed. Even so, persistence of the engineered trait over four generations is a significant result.

Stewart et al. [37] used transgenic and non-transgenic cultivars of *Brassica napus* to determine the fitness effects of engineered insect resistance. The transgenic line was engineered for expression of *Bacillus thuringiensis* (*Bt*) endotoxin. Transgenic and non-transgenic individuals were planted in natural (uncultivated) and cultivated plots and subjected to varying levels of herbivory. The authors found that expression of the transgene conferred a fitness advantage in terms of both increased survivorship and increased reproduction under conditions of insect herbivory. This represents a significant finding, although it should be noted in the context of this discussion that the entities used were not crop-weed hybrids.

Arriola and Ellstrand [35] compared the relative fitness of weed-crop hybrids of *Sorghum* (= *Sorghum halepense* × *S. bicolor*) with nonhybrid johnsongrass (= *S. halepense*). They found no significant differences between hybrid and nonhybrid weeds in several measures of sexual and vegetative reproductive characters (date to first flowering, panicle number per plant, number of seeds per panicle, pollen stainability, number of tillers per plant, total above-ground biomass, and total below-ground biomass). They concluded that the opportunity for hybrid weeds to become established and persist in agricultural settings will be equivalent to that of nonhybrid weeds. In an earlier study, Klinger and Ellstrand [34] measured fitness components in crop-weed hybrids of *Raphanus sativus*. They found that the fitness of hybrids equalled or exceeded weeds for all characters measured (germination success, time to first flowering, fruit production, seed production, and frequency of transmission of the crop allele to seed progeny), and that fruit and seed production were about 15% higher in hybrids than in their wild siblings. They suggested that transgenic hybrids could exhibit similar levels of fitness, allowing transgenes to persist once introduced. Langevin et al. [13] found that weed-crop hybrids of *Oryza sativa* were significantly taller, had more tillers, and larger flag leaves than the related weed red rice. The authors interpreted this as heterosis among first-generation hybrids. The persistence of heterotic characters and their relative contribution to overall fitness was not tested, but the results nevertheless indicate that hybrids suffered no obvious initial decline in fitness over their weedy relatives.

Spread

The spread of transgenes *via* hybridization requires temporal persistence and spatial dispersion of the engineered construct across multiple backcross generations, and will be influenced by multiple factors, including population size, mating system, selective advantage of the introduced trait, and the use of drive mechanisms to sustain the construct in target populations. Unfortunately, direct measures of transgene spread beyond the first generation are almost completely lacking. Nonetheless, the spread of transgenes poses great potential hazard to the environment, because negative impacts will increase in proportion to the area affected. In contrast, genes that escape and persist without spreading pose relatively less environmental hazard, because the area affected will be smaller and the likelihood of successful containment or eradication therefore will be greater.

The spread of transgenes might be predicted through extension of models created to describe the initial probability of gene flow from crops to weeds (e.g. [19, 38]), though such models would need to be modified to account for rates of introgression and the dynamics of small populations. Hybridization and introgression are likely to occur at different rates, and adequate model parameterization will therefore depend on direct, independent estimates of each process.

Estimates of rates of introgression are currently lacking. Further, the dynamic trajectories of very small populations will be highly variable. As noted earlier, risk-averse strategies must account for the average behavior of such models, as well as low-frequency, high-magnitude events.

It is important to recognize that the spread of transgenes through hybridization will be characterized by a different dynamic than direct invasion of the habitat by nonhybrid transgenic plants. Invasiveness of nonhybrid transgenics (reviewed in [39]) will depend on the selective advantage of the transgene and on characteristics of the crop phenotype, while transgene spread *via* hybridization will depend on the selective advantage of the transgene as expressed in a wild-type background.

Containment strategies

Pollen from transgenic crops will be difficult or impossible to contain. Border or trap rows of sacrificial, non-transgenic plantings and barren zones both have been tested for efficacy in pollen containment, with varying results. Umbeck et al. [31] tested the utility of border rows surrounding a plot of genetically engineered cotton. They found that border rows significantly reduced the amount of pollen disseminated from the crop, but the degree of containment varied with compass direction (probably due to variability in pollinator behavior). The authors acknowledged that complete containment is probably impossible. Manasse [26] reported for *Brassica campestris* that "large isolation distance will simply increase mean gene flow, and therefore (will) ultimately increase gene spread". She went on to suggest that border rows might be used to trap pollen leaving the crop. The effectiveness of border rows and barren zones was further investigated by Morris et al. [40] for plantings of *Brassica napus*. These authors found that barren zones were ineffective at preventing gene flow to outlying populations, and might even increase gene flow to distant populations. Conversely, trap rows were effective at reducing gene flow to distant populations, but their effectiveness was highly dependent on the width of the trap. Trap rows 4 m wide were wholly ineffective, whereas trap rows 8 m wide were moderately effective. In no case was gene flow to distant recipients eliminated.

Staniland et al. [41] characterized outcrossing rates within a 30 m wide border surrounding a 30×30 m plot of transgenic *Brassica napus*. They found that outcrossing decayed exponentially across the border area, and concluded that the border area was "very effective" at containing transgenic pollen. However, gene flow to the outer edge of the border ranged from 1–2%, implying that some gene escape likely occurred. Hokanson et al. [28] tested the effects of border rows and trap/donor ratios on gene flow to distant recipients in *Cucumis sativus*. They reported that increasing the trap-donor ratio significantly reduced the movement of pollen to distant recipients. However, in the most effective treatment, the size of the border exceeded the size of the crop itself by 400:1. The authors suggested that borders will only be effective if they are substantially larger than the crop. This configuration seems unlikely to be widely adopted in commercial settings.

Effective containment of gene flow by pollen will not be possible in all agricultural systems, and will be especially difficult in wind-pollinated systems. Indeed, Parker and Bartsch [39] have suggested that "gene flow into wild relatives will occur if it possibly can occur". Risk-averse strategies therefore need to specify acceptable levels of distance-dependent gene flow, and to include containment or other measures that will ensure that specified levels of gene flow are not exceeded. Clearly, more research on the effectiveness of various containment measures are needed, and these will require careful design and application on a case-by-case basis.

Recommendations

The goal of biosafety assessment is the development of risk-averse strategies in the commercial use of transgenic products. The development of effective strategies for the safe use of transgenic agricultural products will depend on adequate biological and ecological characterization of the system(s) of interest that can only be achieved through a combination of appropriate field tests conducted in relevant environment(s) and development of appropriate models. We suggest that the average behavior of models used to predict the relationship between gene flow and distance will generally underestimate the hazard presented by low frequency, large magnitude pollination events, and model selection and parameter estimation should therefore be weighted towards fitting worst-case scenarios at the expense of average model performance. This is especially true for species, cultivars, or phenotypic traits that are inherently hazardous. Further, uncertainty will be created both through variability in biological process and through variability in factor estimation. Risk-averse strategies should seek to incorporate uncertainty from both sources.

We offer the following recommendations based on our review of the literature:

(1)The potential for hybridization has been well-studied in a number of crop-weed systems, but other systems remain relatively unstudied. Therefore, new gene flow studies should seek to measure the potential for hybridization in those systems for which no data exist. In the absence of data to the contrary, one should assume that introgressive hybridization between crops and weeds will occur.

(2)The effects of multiple source populations, repeated introductions, and size of donor and recipient populations on distance-dependent gene flow should be determined. Use of commercial-sized plots is particularly important to accurate estimation of gene flow.

(3)The potential for persistence and spread of transgenic hybrids should be carefully evaluated, using direct measures (such as gene frequency) wherever possible.

(4)Containment measures capable of restricting distance-dependent gene flow to acceptable, specified levels must be developed and carefully evaluated for efficacy.

Finally, we recommend that the current scope of field trials be expanded to address these issues so that risk-averse strategies that incorporate uncertainty can be developed and applied to the problem of transgene movement *via* gene flow.

Acknowledgments
D.P. DeMaster made helpful comments on earlier versions of this manuscript

References

1. Hodgson J, Sugden AM (eds) (1988) *Planned release of genetically engineered organisms. Trends Biotech/Trends Ecol Evol* Special Publication. Elsevier, Cambridge
2. Regal PJ (1988) The adaptive potential of genetically engineered organisms in nature. *In*: Hodgson J, Sugden AM (eds): *Planned release of genetically engineered organisms. Trends Biotech/Trends Ecol Evol* Special Publication. Elsevier, Cambridge, S36–S38
3. Snow AA, Palma PM (1997) Commercialization of transgenic plants: potential ecological risks. *BioScience*

47: 86–96
4. Whitton J, Wolf DE, Arias DM, Snow AA, Rieseberg LH (1997) The persistence of cultivar alleles in wild populations of sunflowers five generations after hybridization. *Theor Appl Genet* 95: 33–40
5. Sossey-Alaoui K, Rajapakse S, Miller MB, Abbott AG, Tonkyn DW, Spira TP (1996) Gene flow from cultivated to wild strawberry (Fragaria spp.). *In*: *Proceedings of the 8th Symposium on Environmental Releases of Biotechnology Products: Risk Assessment Methods and Research Progress.* Ottawa, Canada
6. Arriola PE, Ellstrand NC (1996) Crop-to-weed gene flow in the genus *Sorghum* (Poaceae): spontaneous interspecific hybridization between johnsongrass, *Sorghum halepense*, and crop sorghum, *S. bicolor*. *Amer J Bot* 83: 1153–1160
7. Eber F, Chèvre AM, Baranger A, Vallée P, Tanguy X, Renard M (1994) Spontaneous hybridization between a male-sterile oilseed rape and two weeds. *Theor Appl Genet* 88: 362–368
8. Skogsmyr Io (1994) Gene dispersal from transgenic potatoes to conspecifics: a field trial. *Theor Appl Genet* 88: 770–774
9. Santoni S, Bervillé A (1992) Evidence for gene exchanges between sugar beet (*Beta vulgaris* L.) and wild beets: consequences for transgenic sugar beets. *Plant Mol Biol* 20: 575–577
10. Wilson H, Manhart J (1993) Crop/weed gene flow: *Chenopodium quinoa* Willd. and *C. berlandieri* Moq. *Theor Appl Genet* 86: 642–648
11. Klinger T, Elam DR, Ellstrand NC (1991) Radish as a model system for the study of engineered gene escape rates *via* crop-weed mating. *Conserv Biol* 5: 531–535
12. Klinger T, Arriola PE, Ellstrand NC (1992) Crop-weed hybridization in radish (*Raphanus sativus*): effects of distance and population size. *Amer J Bot* 79: 1431–1435
13. Langevin S, Clay K, Grace JB (1990) The incidence and effects of hybridization between cultivated rice and its related weed red rice (*Oryza sativa* L.). *Evolution* 44: 1000–1008
14. Raybould AF, Gray AJ (1994) Will hybrids of genetically modified crops invade natural communities? *Trends Ecol Evol* 9: 85–89
15. Brown J, Thill DC, Brown AP, Brammer TA, Nair H (1996) Gene transfer between canola (*Brassica napus*) and related weed species. *In*: *Proceedings of the 8th Symposium on Environmental Releases of Biotechnology Products: Risk Assessment Methods and Research Progress.* Ottawa, Canada
16. Jorgensen RB, Andersen B, Landbo L, Mikkelsen TR (1996) Spontaneous hybridization between oilseed rape (*Brassica napus*) and weedy relatives. *Acta Hort* 407: 193–200
17. Mikkelsen TR, Andersen B, Jorgensen RB (1996) The risks of crop transgene spread. *Nature* 380: 31
18. Jorgensen RB, Andersen B (1994) Spontaneous hybridization between oilseed rape (*Brassica napus*) and weedy *B. campestris* (*Brassicaceae*): A risk of growing genetically modified oilseed rape. *Amer J Bot* 81: 1620–1626
19. Kareiva P, Morris W, Jacobi CM (1994) Studying and managing the risk of cross-fertilization between transgenic crops and wild relatives. *Mol Ecol* 3: 15–21
20. Timmons AM, Charters YM, Crawford JW, Burn D, Scott SE, Dubbles SJ, Wilson NJ, Robertson A, O'Brien ET, Squire GR, Wilkinson MJ (1996) Risks from transgenic crops. *Nature* 380: 487
21. Simberloff D (1991) Keystone species and community effects of biological introductions. *In*: L. Ginzburg (ed.): *Assessing ecological risks of biotechnology*. Butterworth-Heinemann, Boston, MA, 1–19
22. Ellstrand NC (1992) Gene flow by pollen: implications for plant conservation genetics. *Oikos* 63: 77–86
23. Ellstrand NC, Elam DR (1993) Population genetic consequences of small population size: implications for plant conservation. *Annu Rev Ecol Syst* 24: 217–242
24. Otto SP, Whitlock MC (1997) The probability of fixation in populations of changing size. *Genetics* 146: 723–733
25. Ellstrand NC, Devlin B, Marshall DL (1989) Gene flow by pollen into small populations: data from experimental and natural stands of wild radish. *Proc Natl Acad Sci USA* 86: 9044–9047
26. Manasse RS (1992) Ecological risks of transgenic plants: effects of spatial dispersion on gene flow. *Ecol Appl* 2: 431–438
27. Goodell K, Elam DR, Nason J, Ellstrand NC (1997) Gene flow among small populations of a self-incompatible plant: an interaction between demography and genetics. *Amer J Bot* 84: 1362–1371
28. Hokanson TC, Grumet R, Hancock JF (1997) Effect of border rows and trap/donor rations on pollen-mediated gene movement. *Ecol Appl* 7: 1075–1081
29. Arias DM, Rieseberg LH (1994) Gene flow between cultivated and wild sunflowers. *Theor Appl Genet* 89: 655–660
30. Till-Bottraud I, Reboud X, Brabant P, Lefranc M, Rherissi B, Vedel F, Darmency H (1992) Outcrossing and hybridization in wild and cultivated foxtail millets: consequences for the release of transgenic crops. *Theor Appl Genet* 83: 940–946
31. Umbeck PF, Barton KA, Nordheim EV, McCarty JC, Parrott WA, Jenkins JN (1991) Degree of pollen dispersal by insects from a field test of genetically engineered cotton. *J Econ Entomol* 84: 1943–1950
32. Metz PLJ, Jacobsen E, Nap JP, Pereira A, Stiekema WJ (1997) The impact on biosafety of the phosphinothricin-tolerance transgene in inter-specific *B. rapa* × *B. napus* hybrids and their successive backcrosses. *Theor Appl Genet* 95: 442–450
33. Chèvre AM, Eber F, Baranger A, Renard M (1997) Gene flow from transgenic crops. *Nature* 389: 924
34. Klinger T, Ellstrand NC (1994) Engineered genes in wild populations: fitness of weed-crop hybrids of *Raphanus sativus*. *Ecol Appl* 4: 117–120
35. Arriola PE, Ellstrand NC (1997) Fitness of interspecific hybrids in the genus *Sorghum*: persistence of crop genes in wild populations. *Ecol Appl* 7: 512–518

36. Linder CR, Schmitt J (1994) Assessing the risks of transgene escape through time and crop-wild hybrid persistence. *Mol Ecol* 3: 23–30
37. Stewart CNJr All JN, Raymer PL, Ramachandran S (1997) Increased fitness of transgenic insecticidal rapeseed under insect selection pressure. *Mol Ecol* 6: 773–779
38. Manasse RS, Kareiva P (1991) Quantifying the spread of recombinant genes and organisms. *In*: L. Ginzburg (ed.): *Assessing ecological risks of biotechnology*. Butterworth-Heinemann, Boston, 214–231
39. Parker IM, Bartsch D (1996) Recent advances in ecological biosafety research on the risks of transgenic plants: A transcontinental perspective. *In*: J. Tomiuk, K. Wohrmann, A. Senkter (eds): *Transgenic Organisms: Biological and Social Implications*. Birkhäuser, Basel, 147–162
40. Morris WF, Kareiva PM, Raymer PL (1994) Do barren zones and pollen traps reduce gene escape from transgenic crops? *Ecol Appl* 4: 157–165
41. Staniland BK, McVetty PBE, Friesen LF, Yarrow S, Thiel P, Freyssinet G, Freyssinet M (1996) Assessing the effectiveness of border areas in confining the spread of transgenic *Brassica napus* pollen. *In*: *Proceedings of the 8th Symposium on Environmental Releases of Biotechnology Products: Risk Assessment Methods and Research Progress*. Ottawa, Canada, 23

Risk assessment of gene flow from a virus-resistant transgenic squash into a wild relative

Marc Fuchs and Dennis Gonsalves

Department of Plant Pathology, Cornell University, New York State Agricultural Experiment Station, Geneva, NY 14456, USA

Introduction

Plant viruses cause important economic losses to agriculture worldwide. In recent years, genetic engineering has provided a major breakthrough for the development of virus-resistant plants. The majority of virus-resistant plants are currently obtained based on the concept of pathogen-derived resistance [1]. The successful use of this technology was first described by Powell-Abel et al. [2] who showed that transgenic tobacco plants expressing the coat protein (CP) gene of tobacco mosaic tobamovirus were protected from infection by this virus. Since this pioneering work, numerous plants, including agronomically important crops, have been engineered for virus resistance, and several field trials have shown the efficiency of this strategy to protect plants from the deleterious effects of plant viruses [3].

Commercial release of virus-resistant transgenic crops has become a reality in the United States. The first virus-resistant transgenic crop, which received exemption status in 1994, was a summer squash resistant to the potyviruses zucchini yellow mosaic virus (ZYMV) and watermelon mosaic virus 2 (WMV 2) [3, 4]. Another transgenic summer squash line, expressing the CP genes of ZYMV, WMV 2 and cucumber mosaic cucumovirus (CMV), and resistant to these three viruses [4], has been recently cleared for commercial use. Also, transgenic papaya expressing the CP gene of papaya ringspot potyvirus and exhibiting resistance to this virus [5], received exemption status for commercial release. More virus-resistant crops are likely to be released in the near future.

Gene flow is a major environmental safety issue for the commercial release of virus-resistant transgenic crops, as well as heterologous encapsidation and recombination. Numerous articles and reviews have been published on potential risks of gene escape from transgenic crops, including virus-resistant transgenic crops [6–10]. However, limited information is available from experiments conducted in the field and none on the establishment of viral transgenes in a free-living plant species yet.

Objectives and rationale of our experiments

We addressed the issue of gene flow by monitoring the dispersal of CP genes from a virus-resistant transgenic squash into a wild relative. Field experiments were designed to estimate the rate of CP gene introgression and to evaluate the fitness of wild x transgenic crop hybrids.

Commercial transgenic squash (*Cucurbita pepo* L. spp. *ovifera* var. *ovifera*) line CZW-3 containing the CP genes of CMV, ZYMV and WMV 2, as well as the marker gene neomycin phosphotransferase (NPT II) [4], was used as the source of transgenic pollen. Its wild relative *C. texana* (*C. pepo* L. spp. *ovifera* var. *texana*), commonly known as Texas gourd, was used as the receptor of transgenic pollen. *C. texana* is a geographically restricted wild growing cucurbit species essentially found in several south-central states in the United States [11]. *C. texana* and transgenic CZW-3 squash are distinctly different in regard to their level of virus resistance, growth habit, fruit characteristics, and seed size. *C. texana* and transgenic CZW-3 squash can readily hybridize without loss of fertility.

Since gene flow has been described between non-transgenic squash cultivars and *C. texana*, and *vice versa* [11], it is reasonable to assume that it will also occur between virus-resistant transgenic squash and *C. texana*, through pollen movement. Thus, we focused our work on the outcomes of an initial hybridization and developed F_1 hybrids (= *C. texana* × CZW-3) by hand pollination in the greenhouse.

Introgression of CP genes from wild × transgenic hybrid squash into wild squash

The first objective of our study was to monitor the dispersal of CP genes from *C. texana* × CZW-3 hybrids into wild *C. texana*, and their subsequent establishment in a population of *C. texana*. Field experiments were conducted over three consecutive years at the same locations, under conditions of high or low disease pressure. The first year hybrids were placed in the center of each field, surrounded by a population of *C. texana*. Thus, the only source of transgenic pollen was from the hybrids. The following 2 years, progenies obtained the previous years were analyzed. The movement of the CP and NPT II transgenes was monitored at the end of each growing season by testing germinating seeds of *C. texana* fruits for the expression of the NPT II protein by ELISA.

Our results showed that CP and NPT II transgenes easily moved from transgenic F_1 hybrids into the surrounding *C. texana* population. Thirty-two percent (846 of 2670) of the *C. texana* offspring tested reacted positively for NPT II in ELISA. Movement of CP and NPT II transgenes continuously occurred over three generations of *C. texana*, providing resistance to CMV, ZYMV, and WMV 2 to the wild squash. However, we found that the rate of transgene introgression was severely inhibited under conditions of high disease pressure.

Fitness evaluation of wild × transgenic hybrid squash

A second objective of our study was to compare the fitness of *C. texana* × CZW-3 hybrids and wild *C. texana*. Plants were tested under conditions of high or low disease pressure.

Results showed that transgenic hybrids (F_1, and first and second back cross generations) that acquired the CP genes exhibited increased fitness over wild *C. texana* under conditions of high disease pressure. Transgenic hybrids displayed resistance to CMV, ZYMV, and WMV 2, produced a higher number of fruits and more viable seeds compared to *C. texana* and non-transgenic hybrid segregants. Under conditions of low disease pressure, transgenic hybrids did not appear to have fitness advantages over their non-transgenic counterparts and *C. texana*, as expected.

Discussion

Our experiments showed the occurrence of gene flow from an agronomically important virus-resistant transgenic crop into a free-living relative, and the subsequent transmission of the virus resistance trait to the wild plants. What are the consequences of the dispersal of CP genes from a commercial virus-resistant transgenic squash into a wild squash? At this point in time, it seems that gene flow is inevitable with transgenic crops, including virus-resistant transgenic crops [7, 8]. A critical question is to assess whether gene flow will enhance the development of undesirable traits such as increased weediness. In other words, will virus resistance confer a selective advantage to *C. texana* in nature? If so, will virus-resistant *C. texana* become a significant threat to the environment as an invasive weed? Or, will viruses limit the ability of *C. texana* to become a weed problem? Extensive surveys would be useful to estimate the incidence of viruses in natural populations of *C. texana*. Interestingly, preliminary information suggests that *C. texana* are not readily infected by viruses in their natural ecosystems. Thus, virus-resistant transgenic squash should not pose undue risks to the environment.

References

1. Sanford JC, Johnston SA (1985) The concept of parasite-derived resistance-deriving resistance genes from the parasite's own genome. *J Theor Biol* 113: 395–405
2. Powell-Abel P, Nelson RS, De B, Hoffmann N, Rogers SG, Fraley RT, Beachy, RN (1986) Delay of disease development in transgenic plants that express the tobacco mosaic virus coat protein gene. *Science* 232: 738–743
3. Fuchs M, Gonsalves D (1997) Genetic Engineering. *In*: Rechcigl NA, Rechcigl JE (eds): *Environmentally Safe Approaches to Crop Disease Control.* Lewis Publishers/CRC Press, Boca Raton, 333–368
4. Tricoli DM, Carney KJ, Russell PF, McMaster JR, Groff DW, Hadden KC, Himmell PT, Hubbard JP, Boeshore ML, Reynolds JF et al. (1995) Field evaluation of transgenic squash containing single or multiple coat protein gene constructs for resistance to cucumber mosaic virus, watermelon mosaic virus 2, and zucchini yellow mosaic virus. *Nature Biotech* (formerly Bio/Technology) 13: 1458–1465
5. Lius S, Manshardt RM, Fitch MMM, Slightom JL, Sanford JC, Gonsalves D (1997) Pathogen-derived resistance provides papaya with effective protection against papaya ringspot virus. *Mol Breeding* 3: 161–168
6. Dale PJ (1992) Spread of engineered genes to wild relatives. *Plant Physiol* 100: 13–15
7. Hancock JF, Grumet R, Hokanson SC (1996) The opportunity of escape of engineered genes from transgenic crops. *Hortscience* 31: 1080–1085
8. Kareiva P, Morris W, Jacobi CM (1994) Studying and managing the risk of cross-fertilization between transgenic crops and wild relatives. *Mol Ecol* 3: 15–21
9. Snow AA, Palma PM (1997) Commercialization of transgenic plants: Potential ecological risks. *Bioscience* 47: 86–96
10. Tepfer M (1993) Viral genes and transgenic plants. What are the potential environmental risks? *Nature Biotech* (formerly Bio/Technology) 11: 1125–1129
11. Wilson HD (1990) Gene flow in squash species. *Bioscience* 40: 449–455

Methods for Risk Assessment of Transgenic Plants
III. Ecological risks and propspects of transgenic plants
K. Ammann, Y. Jacot, V. Simonsen and G. Kjellsson (eds)

RNA recombination in transgenic virus resistant plants

Pia Malnoë, Gàbor Jakab, Eric Droz and Fabian Vaistij

Station Fédérale de Recherches en Production Végétale de Changins, CH-1260 Nyon, Switzerland

Viral infections in crops can cause serious economic losses. The best way to control a viral infection in the field is to use plants that are resistant to the virus. Such plants can be obtained by introducing natural resistance genes by classical breeding methods. However, these genes are often linked to undesirable character traits. Some ten years ago, it was suggested that genetic engineering could be used to increase virus resistance in plants. The idea was based on the observation that inoculation with a non-virulent virus strain conferred protection from infection by a virulent strain. This phenomenon is called *cross protection* and has been used to protect crops in the field. Instead of inoculating the whole virus onto the plant, the insertion of only part of the viral genome into the plant chromosomes was suspected to provide a protection against the virus. This was confirmed and transgenic virus resistant plants expressing a viral sequence are now starting to appear on the market.

In order to obtain an increased resistance to Potato Virus Y (PVY) in several potato varieties, we have cloned the cDNA sequence corresponding to the coat protein gene of this virus. This sequence was introduced into the potato chromosomes using *Agrobacterium tumefaciens* as a vector. After having regenerated transformed plants, we controlled their resistance in the greenhouse. Some transgenic lines showed a very good resistance to PVY. One of these, called Bt6, was tested in the field in 1991/1992.

To study the resistance mechanism, we have been using tobacco instead of potato because it is easier to work with this plant and it also belongs to the *Solanaceae* family. We obtained a transgenic tobacco line (4B5) which is resistant to all PVY strains tested thus far. The 4B5 line was self-pollinated and we analysed 50 transgenic R_1 plants for PVY resistance. The ELISA test, which detects the virus, showed that five of the R_1 plants had lost the resistance although they still contained the transgene. These five plants were studied in more detail by Southern and Northern hybridisations using the PVY coat protein gene as a probe. We found that in these plants the transgene was heavily expressed but in plants that still showed resistance PVY mRNA was not detectable. This was an unexpected result but it is important to realise that *in resistant plants there is very little transgenic viral mRNA present* and furthermore there is evidence that PVY resistance is regulated on a post-transcriptional level.

When investigating the risks associated with the use of viral resistant transgenic plants, two different types of risks were identified. One is heterologous encapsidation and the other is RNA recombination. I would like to focus on RNA recombination which is the only one leading to a permanent change in the viral genome.

When a virus infects a plant that synthesises a transgenic mRNA, the viral replication process can also be initiated on the transgenic mRNA if this molecule contains a polymerase binding site. At a certain moment, the replicase might switch template to the viral genomic RNA, and in this way produce a recombined virus. However, if the replicase binding site is missing from the transgenic mRNA, the replication will start on the viral genomic RNA, then switch to the transgene and finally return to the viral RNA template. Again a recombined virus results. This kind of template switching also happens when two closely related viruses infect the same plant. This is a normal process in nature and a way for the virus to get additional genetic information to ensure its evolution. The environment will determine which combination is the most fit. In transgenic plants, this environment is altered.

We have been studying viral recombination between a transgenic mRNA and a genomic viral RNA. In our model system, we need a double recombination event in order to obtain an infectious recombinant virus. We used a transgenic plant expressing the coat protein gene (hence sensitive) and a virus with a deletion in the same gene. This modified virus cannot migrate upwards in the plant because it cannot be encapsidated. Therefore, a systemic infection of the plant requires a recombination event between the transgenic mRNA and the viral RNA. In order to have enough mRNA available for the recombination event, we have to use a sensitive plant as a host. In a resistant plant, the viral RNA is degraded. Consequently we have to use a plant that has lost its resistance but still contains the coat protein gene. Such a plant will synthesise a large amount of coat protein mRNA which will be translated into the coat protein and encapsidate the mutated virus. This means that a systemic infection can develop in the transgenic plant. To be able to demonstrate that the deleted virus has replaced the missing part of its coat protein gene by recombination, we then need to transfer this infection to a normal untransformed plant which is not able to complement the dysfunctional coat protein. This experiment has now been going on for over 6 months in the greenhouse and we have not yet been able to detect such a recombination event. However, the experiment indicates that a transgenic plant which synthesises a functional coat protein creates a new ecological environment. *Clearly, this means that a mutated virus will be able to survive in a transgenic plant if the transgene can complement the viral mutation.* Although a mutated virus cannot survive when transferred to a normal plant, this still implies that growing transgenic plants on a very large scale modifies the ecological environment. In order to avoid this phenomenon, we only introduced parts of the coat protein gene into the genome of a potato variety called Matilda. The part which includes the 3' end of the RNA polymerase gene and the first bases of the coat protein gene produces a good resistance against the N strain of PVY. This means that one can indeed introduce *only a small part of the coat protein gene into the plant genome and still get a good protection. In these plants no coat protein synthesis is possible and hence there is no risk of heterologous encapsidation.*

Both untransformed and transformed plants can be infected by two different strains of PVY. Although these two strains might show several differences in the amino acid sequence of the coat protein, they are closely related. In this experiment we have been using a transgenic tobacco (R1-28), which synthesises very little transgenic mRNA and is resistant to the PVY-N but not to the PVY-O strain. A co-infection was carried out with the two strains. Unexpectedly, three out of 13 infected plants developed symptoms of a PVY N infection. These results could be explained if a recombination event had happened between the two PVY strains. The sequence of the PVY-N strain which is recognised by the RNAse (described above) could have been exchanged for the corresponding PVY-O sequence which is not recognised by the stipulated RNAse since PVY-O is able to infect the R1-28 plant. This was confirmed by a straightforward polymerase chain reaction (PCR). Primers a and b can detect the N strain, primers c and d, the O strain and the combination of primers a and d a recombined virus. As expected, all the plants were infected with PVY O. Only one plant, which had lost the resistance, was infected with

PVY-N (this plant was then left aside). With the primers a and d, four plants gave a positive result and among these, three showed PVY-N symptoms. To determine whether a second recombination event had taken place, we tested for the presence of an additional part of the N genome by using primers e and f. This region could not be detected in one of the four recombinants, namely the one that did not show any N symptoms. This means that a double recombination event had taken place in this plant.

Among the four different recombination events, two were very close to the 5' end of the coat protein gene and the other two were further downstream. We have looked for recombination between PVY-N and other PVY-O strains, and found the same high frequency of recombination but not at the same sites.

It is important to realise that, in nature, recombination between related viral strains occurs all the time. Such recombination events increase the sequence variability and allow the virus to adapt to a new situation. However, under identical external conditions the viral sequence is remarkably stable. The observation of recombined virus particles in four out of 13 transgenic plants with a double infection indicates a high frequency of recombination. In theory, we do not expect the recombination frequency to be different in an untransformed and a transformed plant. However the external selection pressure is different in the two types of plants. The recombined PVY-N has the advantage of having eliminated the sequence which is recognised by the "specific RNAse" and hence is not degraded in the transgenic plant.

In conclusion, transgenic virus resistant plants create a new ecological environment, and new viruses will appear because there is a strong selection for recombinants. This is most probably also true for "natural" virus resistant plants.

Discussion session 5: Population genetics

Phil Dale
I agree that Terry Klingers results are very much in the category of "nice-to-know information". The crucial questions are: What kind of gene, what kind of crop, what kind of environment would a knowledge of the precise distribution of genes and blips and so on be really crucial information in enabling the regulator to make that decision.

Terrie Klinger
I agree with Phil Dale, and in terms of need to know I would go back to something that Dr. Ammann suggested yesterday. High geneflow systems, especially things that are wind pollinated, will typically carry a higher risk. I think if there are indications that some how sensitive natural wild populations occur in the area, and this goes back to biogeographical information we heard yesterday from Pia Rufener Al Mazyad.

Phil Dale
This is in a way making the value judgment that any geneflow is undesirable and you really need to give some examples of the particular genes and species that you would be concerned about.

Terrie Klinger
I'm not saying that all geneflow is undesirable. Geneflows are going to happen, it has happened and it will continue to happen. In terms of being more precise at this point I would hesitate personally to identify specific genes or specific systems. But I think that should be part of the initial assessment that might be provided to the regulator.

Jim White
Yet we talked yesterday about familiarity with Terry Klingers' and Marc Fuchs' presentations. In the US we have an origin of diversity for squashes and still we are the first country that has released transgenes there. We assume geneflow is going to happen, categorically. We applied the familiarity concept at the time when these transgenic squash lines were deregulated, since traditionally bred virus resistant squashes for zuccini-yellow-watermelon virus and cucumber-mosaic virus were already available and those risks have been acceptable under the tradition of plant breeding.

I would like to ask Pia Malnoë one question. Pia, the transgene, it had the three prime and translated sequences.

Pia Malnoë
It had part of it, not the complete one.

Jim White
Those sequences that have been shown to be potential hotspots for recombination. Have you used a transgene that did not contain the three prime and translated, and the other thing is, your situation is very interesting but obviously your plants are totally resistant to virus infection, which is not a situation for the commercialized lines that are deregulated in the US.

Pia Malnoë
First of all, we have only used the five-prime part of the code protein gene to obtain resistance plants also, but we have not looked into the recombination in those plants and it is true in this case the plant is resistant only to one strain of the homologous strain and not to all PVY strains. But we used this kind of situation, because we had to see it on a short time level, so we had put the situation in such a sense that we could eventually measure recombination. But it is a fact that workers have been looking on recombination in between the transgenic mRNA and the virus in general.

John Adams
I have a comment on Terry Klinger's model describing the dispersal. It occurred to me that you might try borrowing a model from transport planners, because that simple distance decay-function with which you began is a

central ingredient in just about every traffic model which has ever been run. And they find, like you found, that there are residuals below the curve of best fit. And they tend to coincide with interactions between areas of high population.

Christian Damsgaard
It is a comment on familiarity and selection, which is a problematic concept in cases like Opuntia in Australia in the last century.

Jim White
I do not agree: The Opuntia example is categorically different from the squashes in the USA. Opuntia is an introduction to Australia, it is not native. There were no natural pests. Cactus moth was introduced as a biological control agent. Cactus moth was tested to ensure its safety, that it did not feed on native Australian plants. And that's how biological control insects are all tested. That is different. Squashes, the free living squashes, were native to the US. The viruses and the aphid vectors are widely prevalent in the US unlike cactus moth populations.

Jeremy Sweet
Terry Klinger, I believe you said in your talk that fitness of wild crop hybrids was higher than that of the wild crop?

Terrie Klinger
I'm speaking of non-transgenics, but the data from non transgenic trials for the most part have indicated that the fitness of crop-wild hybrids is equal to or sometimes slightly exceeds that of their wild relatives. And that was what I intended to say. I'm talking about the F_1, not about backcrosses, where we have really not much information, its just starting to come out.

Klaus Ammann
It might just be a heterosis effect. And if this would be the general rule that weedy hybrids would be much fitter than their parents, evolution would just drive us crazy when you consider all the new hybridization opportunities given through urbanization. I believe there is much more stability in the biggest genetic experiment mankind has ever made called urbanization, so we are lucky to see that these hybrids, the F_1 generation might be a bit fitter, but then they loose their heterosis effect and eventually find new enemies.

Henry Darmency
Some hybrids also show very low fitness, because they produce few seeds, but at the same time these hybrids are more competitive with respect to the parents. So in fact they can establish themselves in a habitat, but they produce very few seeds, consequently there is not much future for them in the long run.

Terrie Klinger
I do not necessarily disagree with you in general, but there are studies that indicate with radish and sorghum that in fact seed production is higher or at least equivalent. That needs not always be the case, I do not know the generality of that.

Glynis Gliddings
I agree with Terry when she said about geneflow data not conforming to a monotonic declining distribution. I have been looking to 13 quite extensive sets of geneflow data for outcrossing grasses. Each of that 13 sets of data was different from every other one. I'm guessing this estimate of gene flow, given that every time you look you come up with a different answer.

Alan Raybould
Can I suggest you might think about looking at estimates of geneflow from natural populations with processes going on for several generations causing all sorts of conditions over different years. You might get reasonable estimates of geneflow from those, rather than looking at single individual experiments.

Terrie Klinger
That is a very good point. Much of Norman Ellstrands' work and work with his colleagues has been directed at natural populations rather then agricultural systems and we may be of opposing minds with data from that. But it does not need to bother you. The benefits of stability may not be evident, but they may be there.

Klaus Ammann

I think it is really important to determine the natural geneflow and we might go back to simple techniques such as the Dutch-Swiss biogeographical assay in the assessment, also using herbaria. There you have a veritable timemachine built in this method, since hybrids documented in herbaria give us a good picture over recent decades.

Simon Barber

Just several comments. The first one on your modeling on gene dispersal. How do you acccount for insect behavior? I mean, I'm just thinking if you have a beehive somewhere, I imagine that bees tend to behave in a certain way and take the shortest distance to get what they need, and consequently, you may find that this is why you have a huge amount of gene dispersal in one.

A comment about crops. I think most of our crops in the present form are environmental cripples. And I would suggest almost that perhaps most of our weeds in natural systems would perhaps be environmental cripples.

Another comment about using crop species for special products. I think people are considering it. Regulatory agencies consider it. Maybe you can do it if you grow it in a greenhouse and it never gets outside?

Jim White

It is well known that bees once pollinating one plant will tend to stay with that plant and pollinate the same kinds of plants. The first data you presented on Hokema, it was not clear what kinds of other plants or whether there were any plants between those different things. And if there are other plants, bees will skip over those and go to those other plants. And this is why the US prefers border rows of the same kinds of plants, because they are more likely to stop there, because if you have sugar beets and you put weed around them, the bees are likely to skip over the weed and go to other free living sugar beets. So it is better to have the same plants, because of the bees. And I think that's why we talk about how bees travel and in fact, I guess in squash production, they like to travel down the row. Simon could better explain, and Terry mentioned that it's is a food safety issue, but in Canada, the high and low erucic acid canolas are grown. And how can they be separated, and how efficient and effective is that system.

Simon Barber

There is limited production of higher erucic acid rape seed for industrial purposes. It is all grown under contract. So, at the farm level it would not get mixed up with canola quality rape seed. I'm sure that there is potential and it must happen, there will be some pollen movement between these crops. But all canola is tested for erucic acid content. And so there will be some slight contaminations regarding the levels of the erucic acid in the canola crop. If they go above a certain point it is not accepted. And so, thinking of the organic farming debate, I do not think people are going to be able to say we have zero genes getting into this stuff, but maybe there would have to be a limit set. And this would be very similar to what's being going on, I guess, for 60–70 years in the production of certified seed, where you are producing varieties which are distinct in uniform and have to meet certain standards with respect to contamination.

Klaus Ammann

Well, crop to crop geneflow is a problem and we already had some nice demonstrations in front of the House of Parliament stating: "we are against gene smog". But on the other hand, you have to see that organic farmers are threatened already now with hybrid high-yield maize without the transgenes. Again I must warn about the focusing effect. On the contrary, if we seriously defocus, then we have now more precision by having the ways and means of genetic engineering. We now have markers and we are able to follow up what markers are doing.

Jan Carel Zadok

I think genetically speaking there is a very simple answer to that. Take care that the trait expressed in the seed is maternally determined.

Klaus Ammann

In oilseed rape there are already several publications showing that this is done and Pia Rufener Al Mazyad has put together a list of crops where maternal lines have been achieved already and there are many coming up, but to my own dismay, with the exception of alfalfa. There it is more difficult.

Jim White

I doubt that pharmaceutical plants are ever going to be on the table for large scale production. I think, in fact I know, the plants engineered for pharmaceuticals in France are field tested in the US. Field testing pharmaceutical plants already for novel products is already in commercial use. But they will not go through the same commercialization process in the US. They do not have to, because Simon said that for high erucic acid there are contracts. For plastic production or pharmaceuticals they will continue to be under APHIS oversight for field testing. And the containment conditions are much more strict for pharmaceuticals to reduce outcrossing level, for example: For corn the isolation-distance in the US is 200 meters. The minimum standard for the negotiations and containment's for pharmaceuticals is 400 meters. They are not going to be treated as normal field crops, you are not going to be able to buy interleucin canola at the seed store. They are treated as a quite different class of stuff with all oversight and inspections all the time. So, they are not going to be treated like a general release. So I think that is a very good level of safety.

Phil Dale
I just questioned the absoluteness of having a transgene-maternally inherited. If seed from that maternal inherited crop fall to the ground, and you follow perhaps the next year or the following year with a food-crop of the same species, then these plants will potentially be growing together. So I think it eliminates the pollen-problem but it does not eliminate the total problem.

There is a project going on in Canada—oilseed rape to produce hirudin, an anticoagulant substance. With a country of the size of Canada, we thought it would be relatively easy to find a corner that could be dedicated to a few hundred or thousand hectares. And I think that even in Europe where things are more congested, you can often find areas that are not really suitable for growing a food crop, but maybe it would then be possible to genetically isolate it as far as possible; but if you worry about a single pollination over ten km or whatever it is, then your gene is really dangerous and you should not be putting it out. If you are bothered about the outpollination, products of that going into the food chain, then you have got a really nasty protein or whatever it is.

Simon Barber
Just a final comment. Phil Dale is right actually. The people who are producing hirudin in canola had to grow it in intermountain valleys in British Columbia, which is where they do not grow ordinary *Brassica napus*. But you cannot construct neat little pigeonholes in which to put categories of things, because you can always come out with an exception which will not fit there.

Session 6: Decision procedures, harmonisation

Methods for Risk Assessment of Transgenic Plants
III. Ecological risks and propspects of transgenic plants
K. Ammann, Y. Jacot, V. Simonsen and G. Kjellsson (eds)

Transgenic plants and safety regulation

Simon Barber

OECD, Direction de l'Environnement, 2 rue André Pascal, Cédex 16, F-75775 Paris, France

Summary. Harmonisation of regulatory oversight of transgenic plants does not mean that all regulatory agencies must have identical legislation, information requirements, assessment processes and resulting decisions. Rather, it means that they seek to build consensus on the science that they use to assess the safety of the release of transgenic plants. Even though there may be differences in the outcomes of biosafety assessments because of the different environments into which transgenic plants may be released, there is recognition that much information (data) already available are relevant to transgenic plant biosafety assessment. The Organisation of Economic Cooperation Member countries, through the development of Consensus Documents, present this knowledge in a format that can expedite the safety assessment of transgenic plants. These documents can be considered as "mutual recognition of data," and are a means of moving towards harmonised regulatory decision making. To date the OECD Member countries are working on three types of Consensus Document: plant species biologies which focus specifically on centres of origin and diversity, and on related plant species with which that species can hybridise; general trait documents which focus on scientific issues arising from the development of such general traits as coat protein mediated virus resistance; and, specific trait documents which focus on the characteristics of specific genes and the resulting gene products that confer the novel trait to the transgenic plant.

Defining the word "harmonisation" as meaning to be in a state of harmony, begs the question "what is harmony?" The word has similar meanings even when used in different situations, for instance, a collation of material of parallel narratives (e.g. the four Gospels of the New Testament) may be in "harmony," as may the different notes of a piece of music when sounded simultaneously by the different members of a symphony orchestra. What the four Gospels, and the different notes in a piece of music are not, however, is "identical." They are not the same in every detail.

In considering how national regulatory agencies regulate transgenic plants, it is obvious that there are differences. A major difference is the "trigger" used to pull a new plant type into the regulatory process. Some triggers are specific to the technology used in developing the new plant types, for instance some agencies require notification and assessment of any plant type developed using recombinant DNA techniques, while others make "novelty" the trigger, and include novel plant types developed by any means. There are also differences in which national agency is involved in the regulation of transgenic plants, for instance Departments of Environment, Trade or Agriculture may in different countries be responsible for regulations governing transgenic plants and environmental safety. There may also be differences in the type of legal instrument used to regulate transgenic plants, for example the legal instrument may take the form of an "Act" developed specifically for the purpose, a "Regulation" developed under existing legislation or even a guideline that on its own is not a legal instrument.

What, then, is common to international transgenic plant regulation? Apart from maintaining safe laboratory conditions for research with rDNA plants, perhaps the first interest of regulatory agencies worldwide was in the environmental safety of transgenic plants intended for release (experimental testing and eventual use) into the environment. As these new plant types moved from the research and development stage towards commercialisation, food safety and livestock feed safety also became important. Clearly there is harmony among national regulatory agencies with respect to the fundamental elements of safety of transgenic plants that should be considered.

In 1995, the Organisation of Economic Co-operation and Development (OECD) [1] published "Analysis of Information Elements Used in the Assessment of Certain Products of Modern Biotechnology"[1]. This summarised the information required for safety assessment of recombinant DNA products by regulatory agencies in 20 of the OECD Member countries. Results of the survey showed a very high degree of commonality of information requirements, in particular for environmental and agricultural considerations. The questionnaire used in this survey had been developed from the OECD "Recombinant DNA Safety Considerations: Safety considerations for industrial, agricultural and environmental applications of organisms derived by recombinant DNA techniques"[2], the so called "Blue Book" published in 1986. This, the first in a series of OECD publications on rDNA technology and safety, identified scientific criteria for the safe use of rDNA organisms, and also devised a general scientific framework for assessment of rDNA applications in industry, agriculture and the environment. It also emphasised that the safety of rDNA products must be considered on a "case-by-case basis," and that research and development of rDNA products should proceed in a "step-by-step" manner, i.e. moving from the laboratory, to small scale field experiments, then to scale-up for large scale trials and commercialisation, as information to make the relevant decisions on safety were acquired. The safety assessment information elements were broken down into three groups: general and scientific considerations; human health considerations; and, environmental and agricultural considerations.

The common blocks of information clearly identified in the "Blue Book" [2] that are relevant to transgenic plant biosafety, and that have been reiterated at many meetings, workshops and consultations on safety assessment of rDNA products held around the globe, have been incorporated into national regulatory regimes. Although, as already stated, these are not identical, they can be considered to be in harmony. These information requirements may be summarised as follows:

Information about the host plant species

Information about the genetic modification including:

  Information about how the modification was achieved

  Information about all DNA donor organisms

  A complete characterization of the resulting modified host plant

  - description of DNA inserted
  - information of the stability of the insertion(s)
  - description of novel protein(s), expression, and properties

Information about the environment into which the modified plant will be released

  Information about the behaviour of the modified plant in that environment when compared to unmodified counterparts

  Information on interactions with other interacting organisms in that environment:

    Is biology altered resulting in changed environmental interactions?

    Do novel proteins themselves result in different environmental interactions?

In 1992, following in the footsteps of the 1986 "Blue Book," [2] the OECD Group of National Experts in Biotechnology, delegates from the OECD Member countries developing the guiding principles for conducting safe research and biosafety assessment of biotechnology products, published "Safety Considerations for Biotechnology" [3]. This further developed the guiding principles from the "Blue Book" and described criteria and principles for good industrial large-scale practices(GILP) for micro-organisms and cell cultures, and good developmental principles (GDP), guidance for the design of small-scale field research with genetically modified plants and micro-organisms. The Safety Considerations were then further developed with publications for Scale-up: Crop Plants in 1993 [4]; and, Micro-organisms as Biofertilizers in 1995 [5]. A key concept in "Safety Considerations for Biotechnology: Scale-up of Crop Plants" is that of "familiarity." Familiarity is described as:

"The knowledge and experience available for conducting a risk/safety analysis;" (it is emphasised that familiarity must be considered on a case-by-case basis.)

"Familiarity with the crop plant (species), environment, trait and interactions does not determine whether the new combination is either safe or risky. Rather, familiarity with the elements (i.e. case-by-case consideration) of an introduction facilitates a risk/safety analysis;" and

"Where there is sufficient familiarity with the crop plant (species), the new trait and the environment of the proposed scale-up, the risk/safety analysis may be expedited."

In developing a further concept of "familiarity", the 29 Member countries of the OECD, through participation in the Expert Group on Harmonization of Regulatory Oversight in Biotechnology at the OECD, have been working towards the development of "Consensus Documents." These may be considered the collation of relevant information that can be used as "technical tools" to assist in safety based regulatory decision making. While it is acknowledged that there are differences among the Member countries, in the levels of experience gained in regulating transgenic plants, Member countries recognise that duplication can easily occur in their efforts to address environmental safety. These documents, first drafted by a "Lead Country" and then worked on to reach consensus by all Member countries, focus on the science of safety evaluation that is common among the national agencies regulating transgenic plants. Consensus Documents intended for use as technical tools in transgenic plant safety assessment fall into three general categories: plant species biologies; issues arising from "general" trait types introduced into plants; and, information on "specific" traits introduced into plants.

"Plant species biologies" are key to all environmental assessments of transgenic plants. To date two have been published: The Biology of *Brassica napus,* Oilseed Rape [6]; and, The Biology of *Solanum tuberosum,* Potato [7]. Drafts on the biologies of other crop plants (wheat, soybean, and rice) and tree species (Norway Spruce, White Spruce and Poplar) have also been developed and are being reviewed by the Member countries as part of the consensus building process. When consensus on the science is achieved the documents are published. The plant species biologies all follow a similar format, and particular attention is given to centres or origin and diversity, and to the potential of the plant species to hybridise with related plants. The general ecology of the plant species is also described, however, OECD Member countries consider that since the ecology of a plant species, and the presence of related species with which it may hybridise, will vary depending upon the environment in which it is released, a statement on the general pest or weed status of the plant cannot be made. Such a determination will be the responsibility of the agency responsible for the environment into which a transgenic plant is to be released. The plant species biologies follow a modular format, and it is possible to add further relevant information, for instance with respect to food or livestock feed safety of a specific plant species at a later date, if required.

"General trait" Consensus Documents are those dealing with traits such as insect resistance, herbicide tolerance, and virus resistance. To date one general trait Consensus Document has been developed and published, "Consensus Document on General Information Concerning the

Biosafety of Crop Plants Made Virus Resistant through Coat Protein Gene-Mediated Protection" [8]. The document focuses on basic virus characterisation, the expression of viral coat proteins in transgenic plants and scientific biosafety issues related to potential effects of coat protein gene-mediated virus resistance in plants.

"Specific trait" Consensus Documents focus on specific genes and their protein products. Two of these Consensus Documents are under development at the present time, one discussing the genes and gene products that result in transgenic plant tolerance to the herbicide glyphosate, and the other the genes and gene products that result in transgenic plant tolerance to the herbicide phosphinothricin (glufosinate ammonium). These will focus on specific genes and the resulting proteins (enzymes) expressed in the transgenic plant, and their general attributes. They will not deal with the general issues associated with herbicide tolerance. Such issues would be better dealt with under the "general trait" format described above.

The OECD Member countries are striving toward regulatory harmonisation of biosafety assessment procedures for transgenic plants. In applying the guiding principles as laid down in the early OECD rDNA Safety Consideration publications, the Member countries are now developing Consensus Documents as a form of mutual recognition of data, scientific data that can be applied on a case-by-case basis to transgenic plant biosafety assessments where those data are relevant.

References

1. *Analysis of information elements used in the assessment of certain products of modern biotechnology* (1995) OECD, Paris
2. *Recombinant DNA Safety Considerations: Safety considerations for industrial, agricultural and environmental applications of organisms derived by recombinant DNA techniques* (1986) OECD, Paris
3. *Safety Considerations for Biotechnology* (1992) OECD, Paris
4. *Safety Considerations for Biotechnology: Scale-up of crop plants* (1993) OECD, Paris
5. *Safety Considerations for Biotechnology: Scale-up of Micro-organisms as Biofertilizers* (1995) OECD, Paris
6. Consensus Document on the Biology of *Brassica napus* L. (Oilseed Rape), No 7 in a *Series on Harmonization of Regulatory Oversight of Biotechnology* (1977) OECD, Paris
7. Consensus Document on the Biology of *Solanum tuberosum* subsp. *tuberosum* (Potato), No 8 in a *Series on Harmonization of Regulatory Oversight of Biotechnology* (1977) OECD, Paris
9. *Consensus Document on General Information Concerning the Biosafety of Crop Plants made Virus resistant through Coat Protein Gene-mediated Protection* (1996) OECD, Paris

Methods for Risk Assessment of Transgenic Plants
III. Ecological risks and propspects of transgenic plants
K. Ammann, Y. Jacot, V. Simonsen and G. Kjellsson (eds)

Monitoring the impact of releases of genetically modified herbicide tolerant oilseed rape in the UK

Jeremy B. Sweet

National Institute of Agricultural Botany, Huntingdon Road, Cambridge, CB3 0LE, UK

Summary. The impacts of the cultivation of "genetically modified" (GM) herbicide tolerant oilseed rape (*Brassica napus)* on agriculture and the environment are being studied. This paper reports results of monitoring sites for up to 3 years where commercial seed crops and variety trials had been grown. The studies include evaluations of the impacts of the agricultural management of the GM crop as well as the direct effects of the release of the GM crop on the environment.

Introduction

Herbicide tolerance

Risk assessments for the release of genetically modified (i.e. GM) oilseed rape have been largely based on studies of non-transgenic rape or of small scale releases of GM oilseed rape [1]. In 1995, seed production of GM glufosinate ammonium tolerant spring oilseed rape (GMSOSR) commenced in UK at three locations ranging in size from 1–3.5 ha. In addition, Plant Genetic Systems established a 5 ha trial site for testing breeding lines of winter oilseed rape, of which approximately half were genetically modified. In 1996 another GMSOSR seed crop and a further trial area of 7 ha of mostly GM winter oilseed rape (GMWOSR) were established. In addition, in 1995 and 1996 GMSOSR was grown in UK National List Trials at four sites and in 1997 both glufosinate and glyphosate tolerant spring rape was trialled at four sites. In 1996/7 both glufosinate tolerant and glyphosate tolerant winter oilseed rape was grown in 10 trials at various locations in UK.

These crops and trials have provided an opportunity to examine the characteristics of these GM crops and to determine whether they have any impacts that are different from those of non-transgenic crops. They are enabling the risk assessments conducted originally by Plant Genetic Systems, Monsanto and AgrEvo to be verified by studies of several trials at a range of geographic sites grown under normal farming conditions.

Herbicides

The cultivation of herbicide tolerant crops will result in changes in herbicide usage from the currently used selective herbicides to the broad spectrum herbicides (e.g. glufosinate and glyphosate) which the varieties tolerate. In addition, the subsequent volunteers generated by these crops may need different herbicides to eradicate them. These changes in herbicide programmes may have different effects on plant and animal biodiversity in fields and field margins. Trials conducted by NIAB are examining impacts on field margins of glufosinate and glyphosate compared with currently used herbicides. In addition, the establishment of volunteer herbicide tolerant oilseed rape in herbicide treated field margins is also being examined to determine whether feral populations are likely to establish adjacent to herbicide tolerant crops.

Studies of plant diversity in crops subjected to the new herbicide programmes are also planned for the future.

Research objectives

The objective of the NIAB research is to determine the likely agricultural and environmental consequences of the cultivation of herbicide tolerant oilseed rape by studying both the direct impact of large scale releases and the impact of any associated changes in agronomic practices.

Materials and methods

The oilseed rape used in this study consisted of transformed breeding lines, parent lines and F1 varieties developed by Plant Genetic systems, [2]; the transformations consisted of the introduction of a Kanamycin resistance gene, a male sterility gene, a male fertility restorer gene and a glufosinate-ammonium tolerance gene (*Bar* gene).

Comparisons were made with the performance and behaviour of the wide range of non-transformed oilseed rape varieties grown in National List trials.

The sites of the crops were monitored using the methods described by Sweet and Shepperson [3, 4].

Monitoring sites

Monitoring sites were established at four locations in Devon, Lincolnshire and Yorkshire where glufosinate tolerant SOSR seed production crops were grown in 1995 and 1996. A site in Cambridgeshire which has grown successive crops of glufosinate tolerant winter oilseed rape since 1995 is also being monitored. National List trial sites of both spring and winter oilseed rape varieties including glufosinate and glyphosate tolerant varieties in Cambridgeshire are also being monitored. At each site a range of monitoring studies are being conducted [3, 4] and this paper considers the preliminary results of certain studies.

Cross-pollination

At each site, pollination of the nearest oilseed rape crop (usually between 0.5–1 km in distance), oilseed rape volunteers and related cruciferous weeds and hedgerow plants up to a distance of 200 m was studied. Seed samples were collected from plants flowering synchronously with the GM rape and tested for the presence of the herbicide tolerance gene using PCR (see below) or by growing the seed and testing the resultant seedlings for herbicide tolerance.

Detection of herbicide tolerance

Plants were tested destructively for herbicide tolerance by spraying with a 1% dilution in water of Challenge or Liberty herbicide consisting of 150 g/l of glufosinate ammonium. Non-destructive testing was done by placing 1 cm diameter filter paper discs soaked in a 1% solution in water of Challenge or Liberty on to leaves of test plants. Sensitivity in both instances was recorded after 5–6 days.

The presence of the *Bar* gene, which is responsible for the oxidation of the phosphinothrycin produced by the glufosinate ammonium herbicide, was also conducted by amplification by PCR of the products of specific primers for the *Bar* gene [5]. The test was sensitive to a dilution of 1 GM plant sample in 100 non-GM plant samples of equal size. Testing was conducted on samples ranging from single plant samples to samples diluted 1:50. PCR tests were conducted jointly by the Laboratory of the Government Chemist and NIAB.

Seed dispersal

Seed dispersal was recorded at each site by testing volunteers that arose in the field margins, neighbouring fields, tracks and roads traversed by farm machinery associated with the GM crop. The seed bank of GM seed remaining in the soil at each site was assessed after three seasons.

Weediness and Invasiveness of GM oilseed rape

Oilseed rape volunteers can be serious weeds of subsequent crops but are normally controlled by post harvest cultivations and applications of herbicides to emerged volunteers either pre- or post-drilling of the following crop. At each monitoring site the numbers of volunteer rape plants were measured post harvest and in the subsequent crops (mostly wheat). Assessments were made of whether numbers were higher or lower than those of non GM oilseed rape, grown under comparable conditions.

In addition, seed of a breeding line of spring oilseed rape was mixed in equal portions with seed of the same line transformed with the *Bar* gene. In August 1996 the seed was broadcast by hand at a rate of 10 kg/ha into two field margins that had previously received the herbicide treatments described in below. The numbers of GM and non-GM oilseed rape plants was assessed over a 2-year period.

Herbicides used on tolerant oilseed rape: effect on field margins

A study commenced in April 1996 to study the effects of glufosinate, glyphosate and a standard spring oilseed rape herbicide programme on plant populations in two field margin sites, one on the NIAB farm and the other on a farm in Grantchester, Cambridgeshire. Each site was subjected to a series of herbicide treatments to simulate drift or overspraying of field margins as follows:

Treatments per site	Number	Replication	Total Plots
Glufosinate - N, 0.1N, 0.01N	3	3	9
Glyphosate - N, 0.1N, 0.01N	3	3	9
Standard - N, 0.1N, 0.01N	3	3	9
Unsprayed	1	3	3
TOTAL	10		30

N = normal dose rate. Glufosinate = "Liberty" Glyphosate = "Roundup"
Standard programme = cycloxydim (Laser) and benazolin + clopyralid (Benazolox).

Both sites were adjacent to fields in arable rotations but with 3 m buffer areas to reduce the likelihood of farm crop agrochemicals being applied to them that might interfere with the study. The sites were managed as though adjacent to spring oilseed rape crops of both herbicide tolerant and non-tolerant crops. Seed of a similar line of oilseed rape (*Brassica napus)* with and without glufosinate tolerance was scattered onto the field margin treatments in August 1996 to simulate seed dispersed at harvest.

The effects of the herbicide treatments on the botanical composition and establishment of feral rape are being assessed over two seasons after the herbicide treatments. This report describes the results of experiments and assessments made up to June 1997, i.e. one year after the herbicide treatments.

Results

Cross-pollination

The seed crops were isolated from other rape crops by at least 0.5 km and were grown on farms or areas of farm that did not normally cultivate oilseed rape. However, oilseed rape volunteers were detected in close proximity to some seed crops and their seed tested for the presence of the transgene. In addition the nearest margin of the nearest oilseed rape crops was sampled for seed which was tested for the presence of the *Bar* gene. To date, no cross-pollination with other oilseed rape crops has been detected in the several thousand seed samples tested, though tests are still being conducted on samples, and molecular tests are incomplete. Similarly, tests of volunteer and feral rape growing within 200 m of the GM crops have not detected any cross-pollination to date, though tests are not yet complete.

Cruciferous weeds were fairly common at most sites but were successfully eradicated from the GM crops by the glufosinate treatments. Field margin populations of *Sinapis arvensis, Raphanus raphanistrum, Brassica napus, Capsella bursa-pastoris, Alliaria petiolata, Hirschfeldia incana* and *Rorippa nasturtium-aquaticum* (one site) were recorded as flowering at the same time as the GM crops and seed collected from them. Seed was grown and tested for herbicide tolerance or subjected to molecular tests for the *Bar* gene. No herbicide tolerant seed has yet been detected. Populations of cruciferous weeds growing near to the GM crops were revisited in 1996 and 1997 and tested for herbicide tolerance. None has been found yet.

Seed dispersal

Most seed of the GM oilseed rape (OSR) crops was transferred to bags from the combine harvesters and transported in this way, so that spillage was minimised. However, at some sites the harvesters travelled up to 4 km on roads post harvest, shedding seed. Populations of OSR volunteers emerged alongside farm tracks leading from the GMOSR crop fields, in the autumn of 1995 and 1996. However, these populations were very transient and gradually declined during the winter until no plants were detectable in the spring.

The GM spring OSR crops were the last crops to be harvested at all sites in 1995 and 1996. Subsequently, the harvesters remained uncleaned until they harvested winter barley crops the following year. In these barley fields patches of GM oilseed rape volunteers emerged after harvest where the barley seed had flushed out GMOSR remaining in the harvesters and deposited it on the ground. These populations were subsequently eradicated by cultivations and herbicides.

To date no feral populations of GMOSR have established at the sites where GM seed crops and trials were grown in 1995 and 1996. However, the soil seed banks at each site will be tested in 1997 as populations could still establish from residual seed.

Weediness and invasiveness

Volunteer populations of GMOSR were assessed in the crops following all the spring oilseed rape seed crops, two National List spring OSR trials containing GM varieties and the winter GM oilseed rape trials sown in 1995 (Tab. 1). At the Cambridgeshire site 48% of the area was sown with GMWOSR, however only 5 of 77 volunteers (6.5%) were tolerant to glufosinate. At the Lincs 96 and York sites volunteer numbers were associated with seed spillage and unsprayed areas. At the other sites volunteer numbers were very low or non-existent.

Table 1. Oilseed rape volunteers occurring in crops following GM oilseed rape crops and trials

Site Herbicide	Crop and Area Crop	Following Volunteers	No. of Tolerant	No.
Lincs 1995	GMSOSR 1 ha	Sugar beet	4	4
York 1995	GMSOSR 3.5 ha	Winter wheat	43*	43
Devon 1995	GMSOSR 1 ha	Winter wheat and winter barley	0	0
Lincs 1996	GMSOSR 1 ha	Winter wheat	120**	120
Cambs 1995	GMWOSR 2.5 ha WOSR 2.7 ha in mixed plots	Winter wheat	77	5
NIAB 1995	SOSR 2 ha GMSOSR 0.03 ha in trial plots	Winter wheat	0	0

SOSR = Spring oilseed rape; WOSR = Winter oilseed rape; GM = genetically modified for glufosinate tolerance
* 35 of these plants were in one area where broad-leaved weed herbicide had not been applied
**Clump of plants due to spillage of seed when bagging from combine harvester.

Table 2. The number of oilseed rape plants that established in the NIAB field margin site treated with various herbicide treatments

Treatment	Rate	Total No. of oilseed rape plants*	No. of GM oilseed rape plants*
Glufosinate	N	3	0
Glufosinate	0.1N	0	0
Glufosinate	0.01N	0	0
Glyphosate	N	6	1
Glyphosate	0.1N	12	3
Glyphosate	0.01N	1	0
Standard	N	0	0
Standard	0.1N	0	0
Standard	0.01N	0	0
Untreated		1	0

*Number of plants established in 3 × 5 m plots of approximately 1 m width, sown with OSR seed at 10 kg/ha rate.

The numbers of rape plants that established in the two field margin sites treated with various herbicide treatments were assessed in 1997 and the results from the NIAB site are shown in Table 2.

At the second site at Grantchester there was considerable pigeon damage and other grazing of the field margin trial area. Numbers of rape plants establishing were very low and plants were destroyed before herbicide sensitivity tests could be completed.

Herbicide effects

The phytotoxic effects of the herbicides applied to field margin and hedgerow plants were assessed at specific times as follows:

One month post spraying: NIAB and Grantchester

Glyphosate caused considerable chlorosis and necrosis of woody hedgerow, herbaceous and graminaceous species at N and a little at 0.1N rates.
Glufosinate caused considerable necrosis at N rate and less at 0.1N in herbaceous and graminaceous species and some brown spotting on woody hedgerow species.
Standard: considerable necrosis observed at N rate and less at 0.1N in herbaceous and graminaceous species and some brown spotting on woody hedgerow species. Some cruciferous weeds little affected.

3 months after herbicide treatment: NIAB

Untreated plots
the most common species was couch (*Elymus repens)*, followed by bindweed (*Convolvulus arvensis*) and *Poa annua* (annual meadow-grass), then the following species were found at lower frequencies; *Dactylis glomerata* (cocksfoot), yarrow (*Achillea millefolium*), creeping thistle (*Cirsium arvense*), nettles *(Urtica diocia*), cow parsley (*Anthriscus sylvestris)* and hogweed (*Heracleum spondylium*).

Glufosinate treatments
Some areas of bare ground were observed at the N rate, and cow parsley, yarrow, bindweed, nettles, *Poa*, creeping thistle and hogweed were less common than in the untreated plots. *Dactylis* and couch were found in similar frequencies to the untreated plots.

Glyphosate treatments
bare ground was common on the N and 0.1N rate plots, but uncommon on the 0.01N rate plots. Couch was found to be reduced at the N rate but not at lower rates. The other common species found in the untreated plots were found on these plots at much reduced frequencies, but at the N and 0.1N rate many other species not found in the untreated plots were found at low frequencies also, indicating an increase in species diversity. Diversity was not increased on the 0.01N plots. N rate caused defoliation of exposed branches of hawthorn and rose.

Standard treatments
matted grasses were found in high frequencies on the N rate plots, but not at the lower treatment rates. Yarrow was found in higher frequencies at the N and 0.1N rates, and *Poa* was found at higher frequencies at the 0.1N and 0.01N rates than on the untreated plots. The other common species found in the untreated plots were found at similar frequencies at all the treatment rates.

Grantchester

Untreated plots
areas of dead grass were the most frequent ground cover. The most frequent species found were couch and nettles, followed by cow parsley, creeping thistle and barren brome (*Bromus sterilis*) at lower frequencies.

Glufosinate treatments
areas of dead grass were less common than in the untreated plots, and creeping thistle and barren brome were less common at all treatment rates. Couch was found at higher frequencies at the 0.1N and 0.01N rates. *Lolium perenne* (perennial ryegrass) was found to be more common that in the untreated plots on the N rate plots. Nettles and cow parsley were found at similar frequencies to the untreated plots at all treatment rates. No damage to the hedgerow caused by application of the herbicide was visible.

Glyphosate treatments
areas of bare ground rather than dead grass were found at all treatment rates. Barren brome was found to be much less common, being found on one 0.01N rate plot only. Couch and nettles were found to be less common on the N and 0.1N rate plots, but more common on the 0.01N rate plots. Cow parsley and creeping thistle were found at the same frequencies as on the untreated plots. In the N rate plots there was no increase in species diversity and some defoliation of exposed branches of woody hedgerow species.

Standard treatment
bare ground rather than dead grass was found, although only on the N rate plots. Dead grass was found at the same frequency as the untreated plots on the 0.1N and 0.01N rate plots. Couch was found to be more frequent than on the untreated plots at all treatment rates. Creeping thistle and barren brome were found to be less frequent at all treatment rates. Garlic mustard (*Alliaria petiolata*), hedge mustard (*Sisymbrium officinale*), scentless mayweed (*Tripleurospermum maritimum*), *Lolium*, bindweed and prickly sow thistle (*Sonchus asper*) were found to be more frequent than on the untreated plots at all treatment rates. No damage to the hedgerow caused by application of the herbicide was visible.

One year after herbicide treatment: NIAB

Untreated plots
Couch was found to be the most common species, followed by cow parsley, and the following species at lower frequencies: *Holcus lanatus* (Yorkshire fog), bindweed, cleavers (*Galium aparine*), hogweed, yarrow, nettles and creeping thistle.

Glufosinate treatments
couch was found to be much reduced in frequency at all treatment rates. Yarrow, creeping thistle, nettles and cleavers were found in lower frequencies on the 0.1N and 0.01N rate plots. *Holcus* was found in higher frequencies on the N and 0.01N rate plots. Cow parsley, bindweed and hogweed were found at similar frequencies to the untreated plots.

Glyphosate treatments
damage to the hedgerow was found on all N rate plots, the lower branches of hawthorn (*Crataegus monogyna*) being set back in growth or dead. Bare ground was also frequent on the N rate plots. Couch was found to be much less common on all treatment rate plots, however *Holcus* was more common on the 0.1N and 0.01N rate plots. Bindweed and cleavers were found to be more common on the N and 0.01N rate plots, whereas creeping thistle was found to be more common on the N and 0.1N rate plots. Oilseed rape plants established on plots of all treatment rates. The other common species found on the untreated plots were found on these plots at similar frequencies. Many other annual species not found on the untreated plots were found at low frequencies, indicating an increase in species diversity.

Standard treatment
Holcus was found at higher frequencies on all treatment rate plots. Nettles and creeping thistle were less common at all treatment rates. Yarrow was found at a higher frequency on the N rate

plots, but at a lower frequency on the 0.1N and 0.01N plots. Couch was found to be less frequent on the N and 0.01N rate plots. The other common species found on the untreated plots were found on these plots at similar frequencies.

One year after herbicide treatment: Grantchester

Untreated plots
barren brome was the most frequent species found, followed by nettles, cow parsley, *Lolium* and cleavers at lower frequencies.

Glufosinate treatment
barren brome was found to be much less common at all treatment rates. Cleavers and nettles were also found at lower frequencies at all treatment rates. *Lolium* was found to be more common at all treatment rates. Cow parsley was found at similar frequencies to the untreated plots. Many other species not found on the untreated plots were found on all treatment rate plots, indicating an increase in species diversity.

Glyphosate treatments
damage to woody hedgerow species was found on two out of the three N rate plots, the lower branches being dead. Barren brome was less frequent at all treatment rates. Bare ground was common on N and 0.1N rate plots, uncommon on 0.01N plots. Nettles and cleavers were found at lower frequencies on all treatment rate plots. *Lolium* was found at higher frequencies on the 0.1N and 0.01N rate plots. Oilseed rape has established on some plots. More species diversity was found on the N and 0.1N rate plots.

Standard treatment
barren brome and cleavers were found at lower frequencies on all treatment rate plots. Bare ground was frequent on the N rate plots. Oilseed rape established on some of the plots. Nettles were found to be less common at the 0.1N and 0.01N treatment rates, and *Lolium* to be more common at the same rates. More species diversity was found compared to the untreated plots.

Discussion

No pollination of oilseed rape and other crucifers has so far been detected in this study. It is assumed that the GM oilseed rape plants found outside the release field sites arose from dispersed seed, either at drilling or, more likely, post harvest. Given the nature of the releases, ie their comparatively small size, their isolation from other crops and use of land that had no record of growing oilseed rape or that rape had not been grown for at least 10 years previously, it is not surprising that the local incidence of cruciferous weeds was low and that few opportunities for cross-pollination occurred.

Seed dispersal was also restricted by the size of the crops and the nature of the handling post harvest, though spillages and distribution of seed occurred *via* the harvesters at some sites. However, where GM seedlings did occur from dispersed seed their survival was very low due to farm operations on cultivated land and various environmental stresses on uncultivated land e.g. predation, frost etc. No feral/volunteer populations of GM rape have been observed to establish outside the release sites, though areas of seed spillage continue to be monitored.

Volunteer numbers of GMSOSR in the crops which followed were generally low, and usually associated with failures in volunteer control. The incidence in a following crop of sugar beet (4 plants/ha) was surprisingly low. Numbers of SOSR volunteers appearing in wheat crops following National List trials recorded at NIAB and elsewhere have also been very low and this generally indicates that both GM and non-GMSOSR are readily controlled by current farm practices.

Generally winter oilseed rape (WOSR) is more widely grown in the UK and its winter hardiness and biennial character enhance its weediness compared with SOSR. At the Cambridge GMWOSR site 77 WOSR plants were counted in 5 ha. Their distribution pattern suggested that they had not been controlled by the normal arable management of the winter wheat crop that followed. However, the low incidence of GMWOSR in this volunteer population suggested that its weediness was not enhanced by the genetic modifications.

The lower establishment rate of GMSOSR in the field margin trials also suggests that GMSOSR does not have enhanced colonising characters. However, these trials generally had low levels of SOSR establishment and testing for the herbicide tolerance gene was seriously affected by heavy predation at one site, so that little weight should be attached to these results. Trials to study establishment of GMWOSR in field margins are currently underway at NIAB.

The effects of the herbicides on the field margins showed, as anticipated, that the broad spectrum systemic fungicide glyphosate showed the highest levels of phytotoxicity removing perennial species and allowing colonisation by annuals, while glufosinate appeared no more phytotoxic than the currently used herbicides. Establishment of oilseed rape volunteers appeared to be enhanced by the glyphosate treatments, though these results need to be treated with extreme caution because of the generally low levels of establishment. However, it may be prudent to advise farmers to avoid allowing drifts of broad spectrum herbicides into field margins, since they may allow volunteer herbicide tolerant and other GM rape to establish which can then provide sources of contaminant seed and pollen for subsequent rape crops.

The effects of the herbicides used on herbicide tolerant WOSR on field margins is currently being studied at NIAB. In addition, the rate of evolution of multiple herbicide tolerance in adjacent crops and plots is being studied and the weediness, invasiveness and herbicide sensitivity of multiple tolerant plants is also being investigated by workers in UK and France [6]. From these studies it will be possible to determine both the agronomic and environmental impacts of herbicide tolerant oilseed rape. NIAB is studying the effects on farming operations and management and the longer term consequences for agriculture.

Acknowledgements
The authors thank the Department of the Environment, Transport and the Regions, AgrEvo and MAFF for financial assistance, and Plant Genetic Systems for kindly providing diagnostic primers, seed and access to trials. Technical assistance was provided by Euan Simpson. The molecular diagnostics of plant samples was conducted by Nigel Burns of the Laboratory of the Government Chemist and Jane Byrne, a Bath University student.

References

1. Harding K, Harris PS (1995) *Risk assessments of the release of genetically modified plants: a review.* MAFF 54 pp
2. Rudelsheim P, Huybrechts I (1995) A case study of hybrid oilseed rape: from conception to marketing. Report of the 4th NIAB International Forum: *Genetically Modified Crop Cultivars.*, 10 pp
3. Sweet JB, Shepperson R (1996) Monitoring commercial releases of genetically modified oilseed rape. *Proceedings of 10th International Weed Biology Conference*, Dijon, 1996: 217–222
4. Sweet JB, Shepperson R, Thomas JE, Simpson E (1997) The impact of releases of genetically modified herbicide tolerant oilseed rape in the UK. *Proceedings of the Brighton Conference*, 1997, 291–302
5. Rogers HJ, Matharu B, Parkes HC (1996) Monitoring releases of transgenic plants: theoretical and practical considerations. *BCPC Symposium Proceedings* 65: 39–45
6. Messean A (1997) Management of herbicide tolerant crops in Europe. *Proceedings of the 1997 Brighton Conference* 947–954

Views of non-governmental organizations on the risk evaluation of genetically modified organisms

Piet Schenkelaars

Schenkelaars Biotechnology Consultancy, Bio Science Park, Niels Bohrweg 11–13, NL-2333 CA Leiden, The Netherlands

Background

The environmental organization *Stichting Natuur en Milieu* was commissioned by the Netherlands' Ministry of the Environment to prepare an inventory of views of non-governmental organizations (i.e. NGOs) on the risk evaluation of genetically modified organisms (i.e. GMOs) in six cases: two cases of the contained use of genetically modified micro-organisms, three cases of field trials with genetically modified plants, and one case of a environmental release of a genetically modified micro-organism. Subsequently, SNM subcontracted Schuttelaar & Partners to prepare the inventory. Five NGOs from the Netherlands, one from Germany and one from the United Kingdom were prepared to participate.

General findings

Viewpoints of NGOs in the Netherlands

One of the basic problems encountered by all NGOs was the high level of technical expertise required to participate. In general, the NGOs perceived a lack of information on the accuracy of the insertion, the number of copies inserted, the sites of integration and on the levels of gene expression and, where appropriate, their developmental and tissue specificity. This information was viewed necessary to assess the risk of increased genomic instability with unknown ecological effects.

Further, all NGOs criticized the vagueness of criteria applied in the consent procedure to establish whether the GMO deviated, except for the modification intended, from the host organism.

Moreover, all NGOs perceived a lack of empirical data on the actual, ecological performance of GMOs, such as their weediness and invasiveness in the case of genetically modified plants compared to that of their non-genetically modified counterparts. The view was shared that the potential long-term ecological effects of their use had not been adequately addressed. Some also explicitly questioned the short-term proposed by the applicant or required by the Competent Authority (CA) to monitor ecological effects resulting from interactions between GMOs and the environment. It was also pointed out that the endpoint of risk analysis determines which parameters are considered to be relevant to monitor, implying that disagreement about the endpoint of risk analysis can lead to disagreement about which parameters should be monitored.

In addition, one NGO provided an extensive, historical review of developments of scientific approaches on the risk assessment of genetically modified organisms. This review attempted to explain why this NGO perceives an inadequate input of up-to-date ecological insights into the decision-making on the use of GMOs. On this basis, it was argued in several instances that the precautionary principle had not been adequately applied in present regulatory approaches to the use of GMOs.

Further, taking as a starting point the classical risk paradigm, *risk = exposure * hazard*, several NGOs concluded that the present approach to risk assessment of GMOs by the CA was confined to an "armchair" hazard identification and a qualitative estimate of "exposure" in terms of the probabilities of such hazards occurring. These NGOs therefore suggested approaches seeking a more empirical science-based and quantified risk assessment of GMOs, in parallel to the risk assessment of chemical substances. For instance, instead of assuming whether a genetic insert would confer a selective advantage to a GMO or recipient organisms, it was proposed to measure the insert's fitness, dispersal and introgression in relation to the connectivity between the GMOs and recipient populations. Similarly, empirical data on the weediness and invasiveness potential of a genetically modified crop plants in comparison to host crop plants should be collected, instead of assuming that the genetic modification would have no impact on these characteristics.

Such empirical information would contribute to a better founding of risk management procedures, to unambiguous requirements regarding data and monitoring, and to an appropriate enforcement of regulations.

Finally, all NGOs addressed the issue that a risk analysis should be distinguished from a risk evaluation. The acceptance of a risk is yet another issue. The NGOs indicated to have encountered the difficulty to identify which data requirements had to be met by the applicants to obtain consent. The NGOs also pointed out that the normative frame of reference for the risk analysis was applied very implicitly in the risk evaluation by the CA. According to the NGOs, every risk evaluation is value-laden. In this case, it was for instance felt that in the risk evaluations under consideration criteria relating to sustainability had not been applied. For this reason, the NGOs concluded that decisions by the CA may be based upon an implicit environmental ethics conflicting with other views on the acceptability of risks involved in the use of GMOs.

Comments by a NGO from the United Kingdom

In its comments on the views of the Dutch NGOs, the Green Alliance from the United Kingdom raised the question about the regulatory handling of uncertainty, as it would be unlikely that there will be ever enough scientific data to make accurate predictions of the long-term ecological impacts of releasing GMOs. In the view of the Green Alliance, the precautionary approach means more than applying "state-of-the-art" ecology; the precautionary approach would also need to take a view about uncertainty, and decide not to consent a release if there are a number of unknowns involved.

Since the evaluation of the risks of a GMO-application was viewed by the NGOs as something different than their acceptance, the Green Alliance pointed out that, at present, the NGOs have not yet got to the stage of indicating the kind of answers that would lead to conclusions about acceptability. For instance, if more (empirical) data on the weediness potential of a genetically modified beet were available, what would be an acceptable level of weediness? Or, in wanting to know the host range of the *Bt* toxin in the potatoes, which potential hosts would be of concern? To the opinion of the Green Alliance, it should be the role of NGOs to set the agenda in terms of what matters about the possible ecological consequences of GMOs, although it recognized that this would be very hard to do.

In addition, the Dutch NGOs, in common with their UK counterparts, raised broad questions about the effects of herbicide tolerance, virus resistance and *Bt* plants as strategies towards sustainable agriculture and addressed the idea of weighing risks and benefits. To the opinion of the Green Alliance, consideration of benefits was not within the scope of Directive 90/220, as the system set up throughout Europe was geared to consideration and judgment of risks. However, as the Dutch NGOs and the Green Alliance observed, there tended to be buried value judgments in the regulatory systems. These included a general value judgment that the technology was worthwhile, and that only a particular level of demonstrable risk, rather than a particular level of uncertainty, justified turning down the application. According to the Green Alliance, a policy mechanisms that sought to weigh up the risks and benefits of the technology, and even alternatives to the technology, could seek to challenge these value judgments.

Comments from a NGO from Germany

BUND/Friends of the Earth-Germany pointed out that hardly any NGO had the scientific expertise and financial resources required to thoroughly evaluate the risks of GMOs. When, due to this limited scientific expertise, the NGOs do not adequately demonstrate the risks involved, the false impression might emerge that the GMO is acceptable by the NGOs. In addition, even if no risks could be demonstrated, there might still be political or ethical considerations not to accept such a GMO.

Second, the implications of a positive evaluation and acceptance of a single product containing or based on GMOs by a NGO should not be underestimated. Such a "door-opener" might be of interest to the company wishing to market that product. However, NGOs, known to oppose gene technology in general, would have difficulties to convey such a "differentiated" position on a single product, and, in addition, risk their public credibility.

Finally, BUND remarked that previous experiments with Technology Assessment (TA) had indicated that a "problem-oriented" instead of a "technology-induced" approach should be preferred. The first step is the problem of how to produce sufficient healthy and nutritious food in an ecologically sustainable way. The second step would then be to evaluate the potential contribution of different alternative solutions to this problem, followed by a third step, in which the potential risks and benefits involved in the most promising technology should be evaluated. Subsequently, the risks and benefits involved in the second best technology should be evaluated. Finally, the risk potentials of both technologies should be compared before reaching a final conclusion.

Risk assessment of transgenic plants – a comparison with pesticide regulation

Werner Mueller[1], Helge Torgersen[2] and Helmut Gaugitsch[3]

[1] *Institute of Organic Farming, University of Agricultural Sciences (BOKU), Gregor-Mendelstr. 33, A-1180 Vienna, Austria*
[2] *Institute of Technology Assessment, Austrian Academy of Sciences, Postgasse 7/4/3, A-1010 Vienna, Austria*
3*Austrian Federal Environment Agency, Spittelauerlände 5, A-1090 Vienna, Austria*

The EU-Directive 90/220/EEC [1] regulating the deliberate release of transgenic organisms contains a list of criteria and in some cases assessment endpoints to consider in the risk assessment of Genetically Modified Organisms (i.e. GMOs). We compared this list with EU- Directive 91/414/EEC [1] regulating risk assessment of chemical and microbiological pesticides, in order to analyse differences between both directives and to find out where comparable experience with conventional organisms and appropriate test methods is available. For this purpose it is helpful to distinguish between different (albeit interlinked) categories of risks that are discussed in the following: "genetic" risks, risks from the gene product, risks with regard to adverse agrosystem effects, and socio-economic risks.

"Genetic" risks result from the dissemination of inserted genes in a new gene pool and from the unwanted survival, dissemination and propagation of GMOs in ecosystems. Directive 90/220/EEC explicitly addresses aspects of this risk category. Although the main focus of Directive 91/414/EEC is on the gene product or chemical risk category (see below) there are many common criteria for risk assessment in both directives such as: pathogenicity, survival, dissemination and propagation of the released organisms, genetic stability, impacts on the food chain and to non-target organisms. There is little depth as far as methods and assessment endpoints are concerned. No threshold values are given for this risk category with the exception of the persistence of chemical and microbiological pesticides in the soil in Directive 91/414/EEC (DT90 < 1 year, DT50 < 3 months).

Although there are some test methods available for transgenic plants [1, 2] there is still a lack of generally accepted methods for adequate risk prediction on a case-by-case basis in this risk category. This mirrors the gap in knowledge on the principles of evolution. Beyond that there is no clear definition of which potential effects on ecosystems should be considered as harmful.

Risks from gene products are assessed by measuring stability, toxicology and ecotoxicology of GMOs and their products. There is a lot of experience from the assessment of environmental and health risks of exposure to chemicals. Knowledge gaps still exist with respect to immunotoxicity and here especially the determination of allergenicity [5], estrogenic [6] and synergistic effects. Contrary to Directive 90/220/EEC, Directive 91/414/EEC, provides detailed definitions of assessment endpoints, evaluation principles and threshold values in this risk category. However, the large amount of methods used for the assessment of chemicals cannot be directly used for assessing gene product risks of GMOs because methods for assessing chemicals have been developed largely for assessing single substances only. Whole food studies to assess the toxicology of organisms and their products meet serious methodological problems (especially with unbalanced diets) [3, 7].

Adverse agrosystem effects are harmful effects (e.g. increased pesticide use) to the environment which arise from the interdependence of new traits and changes in agricultural practice or changes in consumer food preference. Obviously these effects cannot be reduced to a single cause. Studying the ecological impacts resulting from the cultivation of conventional non-transgenic crops, we came to the conclusion that, following the catalogue of questions of Directive 90/220/EEC, it is hardly possible to infer the observed ecological effects directly from the plants′ traits [8]. Both directives do not refer to this risk category, but we propose to include questions on adverse system effects into both directives.

Socio-economic risks are taken into account neither by Directive 91/414/EEC nor 90/220/EEC. Although modelling can be useful for their assessment, neither endpoints nor a common understanding of unwanted effects or developments are established. Nevertheless, some efforts have been made in the genetic engineering laws in Norway and Austria to deal with this category. However, no substantial results have been obtained so far.

Besides these risk categories Directive 91/414/EEC takes into account regional aspects, risk-benefit analyses, and agricultural effects like the development of resistant pests. This standard of pesticide risk assessment should be adopted for the assessment of GMOs as well, with a reassessment after a certain period of time in order to take into account scientific progress. The recent proposal for an amendment of Directive 90/220/EEC by the EU Commission takes up some of these points.

Experience from the history of pesticide regulation

Studying the history of risk identification and regulation of different pesticides shows that from 1939 up till now the risk identification of new (previously unknown) harmful effects to humans and the environment continues to evolve. The depletion of the ozone layer by Methyl-Bromide and the estrogenic effects of many chemicals (for example Vinclozolin, see Fig. 1) are well-known examples of these new risk identifications within the last two decades. Furthermore the history of pesticides regulation shows the time lag between chemical exposure of the environment, the occurrence of a harmful effect, the identification of that deleterious effect, the scientific proof of the cause of that effect and the political reaction to that new scientific evidence. Figure 1 gives a short review of these aspects.

Another lesson from this historical review is that persistent chemicals - especially if they are mobile and can be accumulated – are a major risk factor in a continuously changing environment. Due to the complexity of ecosystems the overall ecotoxicity never can be fully assessed. The persistence of chemicals is a central criteria for assessing the ecotoxicity because the exposure of persistent chemicals cannot be terminated or removed if new harmful effects will be identified in the future [9, 10]. Since GMOs are viable their persistence may be even more critical than in the case of chemical substances. This has to be taken into account in the risk assessment.

	DDT	Methyl Bromide	Vinclozolin
1940	*approval (before 1948)* resistant insects (1947)		
1950	DDT in mother´s milk (1950) accumulation in soil and fat tissue (1951)		
1960		*approval (1965)*	
1970			
1980			*approval (1984)*
1990	approval withdrawn (1992) estrogenic effect (1995)	depletion of the ozone layer (1994)	estrogenic effects (1994)
2000		*phase out (2005)?*	
2010			

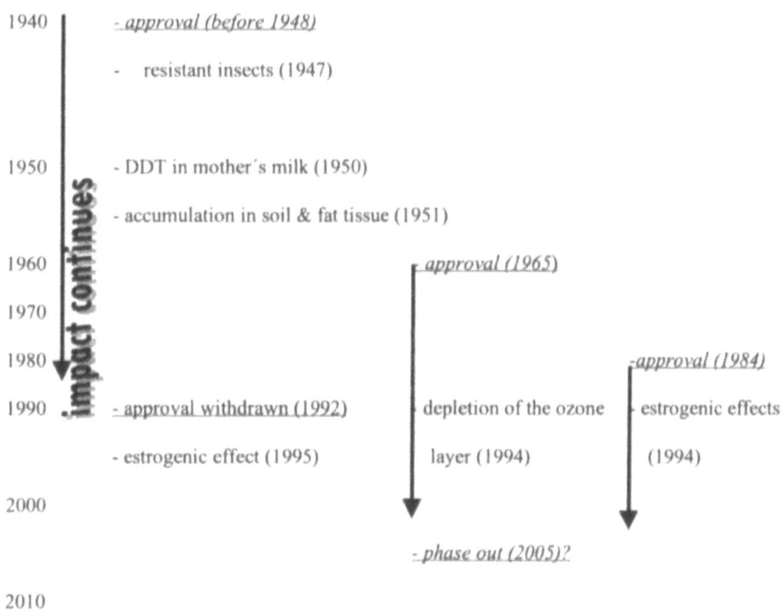

Figure 1. Historical review of the risk identification and regulation of three pesticides in Austria

Conclusions

Comparing EU-Directive 90/220/EEC regulating the deliberate release of transgenic organisms and Directive 91/414/EEC regulating the approval of pesticides reveals great differences in the depth of regulation. This may lead to the paradoxical situation that, for example, a genetically modified plant producing a microbial pesticide protein may undergo a risk assessment (following Directive 90/220/EEC) under less detailed conditions more open to interpretations than the protein or the microbe itself if applied as a pesticide.

The history of pesticide regulation reveals that one of the important questions is how scientific progress in risk identification can be taken into account in the risk assessment. It is obvious that persistence is a central assessment criteria to refer to this problem. The consequences to be drawn from the history of environmental risk assessment of pesticides is the necessity to strengthen the precautionary principle" in the risk assessment procedure not only for pesticides but also for GMOs. Directive 91/414/EEC refers to this fact requiring a limited persistence for chemical and microbiological pesticides. Gene escape of synthetic genes into a new gene pool could be compared with the persistence of chemicals. No thresholds are set for gene escape by Directive 90/220/EEC, which has to be considered a major omission from a precautionary point of view.

References

1. EU-Directive 90/220/EEC (1990) Council Directive of 3. April 1990 on the deliberate release into the environment of genetically modified organism (OJ N° L 117/15, Brussels)
2. EU-Directive 91/414/EEC (1991) Council Directive of 15. July 1991 concerning the placing of plant-protection products on the market (OJ N° L230, 19.8.1991, Brussels)
3. Kjellsson G, Simonsen V (1994) *Methods for risk assessment of transgenic plants. i. competition, establishment and ecosystem effects.* Birkhäuser, Basel
4. Kjellsson G, Simonsen V, Ammann K (eds) (1997) *Methods for risk assessment of transgenic plants. ii. pollination, gene-transfer and population impacts.* Birkhäuser, Basel
5. Kok EJ, Kuiper HA (1996) Evaluation of strategies for food assessment of genetically modified agricultural products – information needs. *In*: *OECD: Food Safety Evaluation*, 80–84
6. Seibert H (1997) Umweltchemikalien mit endokriner Wirkung. *In*: *UBA Umweltchemikalien mit hormoneller Wirkung*. Tagungsberichte Band 19, Wien
7. Hammond B, Rodger SG, Fuchs RL (1996) Limitations of whole food feeding studies in food safety assessment. *In*: *OECD Food Safety Evaluation*, 85–97
8. Torgersen H (1996) Ecological impacts of *Traditional crop plants – a basis for the assessment of transgenic plants? austrian federal environment agency.* Umweltbundesamt, Monograph Nr.75
9. Klöpffer W (1994) Environmental hazard – assessment of chemicals and products part ii: persistence and degradability of organic chemicals. *ESPR – Environ Sci Pollut Res* 1(2): 108–116
10. Mathes K (1996) Langzeitbeobachtung und –bewertung von Umweltveränderungen transgener Organismen. Was hat die Ökologieforschung gebracht? *In*: *Texte 58/96*, Umweltbundesamt, Berlin, 61–70

Discussion session 6: Decision procedures, harmonisation

Jan Husby
The purpose of the Norwegian Gene technology act is to ensure that the production and use of GMO takes place in an ethically and socially justifiably way in accordance with the principle of sustainable development and without detrimental effects on health and environment. Living in a democratic society, we are sending out applications on marketing to a broad public, both consumer organizations, green organizations and scientific institutions and also the industry has the possibility to give their opinions. However, the Norwegian parliament in complete consensus prohibit GMO and products with antibiotic-resistance-genes. Why did the parliament do this? First of all, they used the precautionary principle, because the scientific world did not convince them of the safety of antibiotic-resistance-genes in plants. Quite many experts working within gene technology method in veterinary and medicine research were much more sceptical and had seen the development of antibiotic-resistance pathogen bacteria compared to plant breeders working in virgin technology on plants. And the last question of uncertainty was in fact, do we need them? Do they compete with the traditional not modified plants, are they better for the consumer, are they cheaper? In this respect the biotech industry has not managed to convince the consumers nor the Norwegian politicians and not many of the experts in related fields in Norway. The big issue is the overall lack of knowledge and the uncertainty of positive and negative effects on both, human health and environment, are very important.

Phil Dale
In all aspects of life there are all kinds of uncertainties. Is it fair to say that we take much more notice of risk with transgenics, because it is novel? Piet Schenkelars said that many of the interest groups felt that risk assessment is armchair exercise. The alternative to an armchair exercise is to get out and to do field releases and to commercialize and to learn from that. Now, if we did so without doing this so called armchair exercise the interest groups would be the first to criticize that we will not be drawing on all of the experience and data that was available. One assumption is that science gives black and white answers, but we all know that they are usually shades of grey, because we do not know everything about everything, we have to try to make decisions in using all information we can. The other point about risk assessment being poorly integrated seeming a bit of a model. It does seem confused and confusing when we talk about things at generic level, when we talk about gene-transfer being a hazard, when we talk about herbicide tolerance being a hazard. You can only really get down to think about particular issues, solutions by talking about specifics, by talking about particular genes, particular crops and particular environments. The risk assessment is very much built on science, but there is a political element in the sense that a country has to make a decision if a particular area of technology is worth pursuing and has some value. One of the problems in decision making is, if you have wide extremes in a group of people trying to decide which are safe and which are unsafe. If you have people that want to stop the whole thing and people that want to move along as quickly as possible without any regulation or with minimal regulation then this resembles a debate of fachists and communists about democracy. And eventually they have to make a decision with all of the uncertainties that are involved.

Les Levidow
Simon Barber described that the OECD has to develop at least criteria for data requirements and the future acceptance to data and in that respect the criteria are now clearer, but as we all know even data which are mutually accepted does not translate into accepted decisions. A great task is to clarify the uncertainties in every aspect of risk assessment including monitoring after commercial approval. All of this would be a great departure for the European Science Foundation (ESF), at least regarding biotechnology. In EU, DG XI brings together all the competent authorities and they respond to marketing applications in particular that brings out their differences whether they accept the data in the application, how they interpret the data and whether they disagree about how to define an adverse effect under the directive, they disagree about the burden of evidence, demonstrating that a potential effect will not happen and so on. So already there is a clarification process built into the regulatory procedure. Likewise DG XII has substantial funding for the prenormative research program which aims are to fill gaps in knowledge needed by the risk assessment procedure taking place in DG XI. Here, very little research is funded on transgenic plants. We see very few applications for such research. So I have a main question from all this. Why is

there a gap in the institutions which are officially responsible for carrying out the functions of decision making and harmonization and so on?

Jeremy Sweet
My feeling about the European programs is, first of all, they tend to be very narrow and specific and they will look up to particular narrow issue and problem and tackle it probably from a very narrow perspective. Another perspective is to try to get a multi-disciplinary approach to risk assessments. In Europe, we should be considering the agricultural impact in agriculture management. The other reason for going to ESF is that it includes countries not in the EU, but the EFS is specifically supported by the scientific academies of a number of non EU countries as well, for example, Switzerland.

Jan Carel Zadok
Piet Schekelaars gave an excellent overview of what was ongoing in the Netherlands. One point which he brought forward is coming from the English NGO which was apparently a buried value judgement that technology is worthwhile. I do not think this view is correct. It is what the government wants, what parliament wants. It wants to stimulate new technology, but with sufficient safety.

Jeremy Sweet
I have an open mind on the need for monitoring and I think it will be one of the things the ESF group needs to discuss. They need to discuss what you can monitor and why and if they come up with good reasons and so on, then OECD shall develop the methodology. There is a major monitor program in the UK. It is looking at oilseed rape and looking to see if the risk assessment done by the original companies is OK when the oilseed rape is grown on a large scale. What we really need are clear objectives.

Christian Damgaard
Many of the decision makers this afternoon said that risk assessment was a probability, or they did not use the word probability. Probability times effects, that is the risk assessment. And you multiply two things by multiplying two numbers and maybe the numbers are what we actually want and that is actually the question. What I would like is to suggest is that the familiarity concept as used now is based on a terminology which is not used in the science community. I read your booklet on the definition on the familiarity, it pretty much corresponded to what I would call a probability. The advantage of probability is firstly that it is possible to be multiplied with the effect, because it is a number, it is between zero and one, it will also raise a specific question which the scientific community could answer. What is the probability of a certain event? The answer in many circumstances will be very difficult to find, but I think if you ask the question specifically the chance to get a good answer might be larger. Using the word probability instead of familiarity will make the dialogue between scientists and regulators easier.

Simon Barber
The concept of familiarity includes additional things, they probably would not be covered by just by looking at probabilities of events. Many of the familiarity components will not translate easily into probability, such as, for example, genetic variability of oilseed rape. What is probability for something like the potential for a *Brassica* to outcross to another species? We do need to be clearer on the components of the familiarity knowledge base.

Jan Husby
Another dimension in the discussion is the time scale. We are in fact a big and a small audience at the same time. We know the decision procedures we have to follow, we would like to have the same decision procedures, but we know that this is going to be different in the future. With the case of transgenic spruce, which might soon arrive, my grandchildren will have completely different management regimes and systems.

John Adams
The title of my paper tomorrow is a *Richter scale* for risk, but I'm not going to have much time to talk about that. The example I use in my paper is the risk of a road accident or dying in a road accident. The appeal for a Richter scale of risk is based on the idea that you fill the scale benchmark risks. The risk of dying in a road accident is 1:16000 (men/year). What on earth did this mean? If you go through the literature you find that a young man is about a hundred times more likely to die in a road accident than a middle-aged woman. If you are on the road at three o'clock on a Sunday morning you are over a hundred times more likely to die than if you are on the road at ten o'clock on a Sunday morning. If you are two and a half-times over the alcohol limit, it is 20 times. Now *if* all those factors were independent, then a young drunken man on the road at three o'clock on a Sunday morning

would be about 2.5 million times more likely to die than a normal middle-aged woman driving to church seven hours later.

Piet Schenkelaars
Reply to Phil Dales comment: We are living somehow in a democratic society, that does not mean a perfect system, but I do not know a better one. One of my observations is that membership of political parties is declining. So, when political parties decide that the technology is useful, that maybe an interesting thing. On the other hand, membership of consumer and environmental and nature protection organizations are very large in comparison with political parties. Sorry for people of industry, politics and science, but your credibility is quite low in comparison to that of the NGO's. When you want to introduce biotech on the market and you have opinions from NGO's that technology is not worthwhile to pursue, you have a problem. Risk assessment research gets funded perhaps a little bit, maybe not enough. Then I would like to comment that the NGO's at least admit that they have low level of expertise on biotechnology and I know that some NGO's do not want to increase their knowledge base, but there are quite a lot of people who would like to increase the knowledge base and would be enabled to do so. But when you then talk about increasing knowledge within NGO's community programs and communication etc., they are quite well funded. On the other hand I would say that biotechnologists, they are of course experts in biotechnology, but are they experts when you have these very general discussions, are they experts in world economy and how to feed the world? They have also a very specialized scientific expertise and often they stretch their range of expertise in a way they should not do. What I would like to say is, that we have to communicate. And when I think about world communication that is a two-way street.

Klaus Ammann
Remember my term *symmetry of ignorance*, and remember my term step-by-step proceeding, learning from each other in a very pragmatic way, and I think this also should be incorporated in decision processes. Biotech companies have to go through the bottleneck of a very stringent regulatory procedure, when they make a single small fault, they will be just slapped on their wrists. When NGO's are seriously involved, they should be heard, but also they have to behave according to a certain code of conduct.

Phil Dale
I agree with what Piet said about communication and involvement, and as scientists we have an increasing responsibility to do this and to spend time. I did the first release just over 10 years ago and we have been faced with the same sort of question time after time.

The other thing I find with some of the interest groups, like Greenpeace and Friends of the Earth. When you get them in the bar and you can have a proper reasoned debate with them, they will acknowledge that there is potential to do some really useful important things. But when you get them back in the conference room, they spring back on a bit of elastic. They follow the party line, we must stop all of this, it is unsafe, it is unacceptable. I think that is a part of the problem. But clearly, we have to keep humming away at the communication through involvement and information. I think we need to do more in schools, more with consumer groups and so on. I would very much like to get a transgenic plant system in school. We should start to think about that. Together an agrobacterium with a construct approves the regulatory process, there are risk assessments on all the possible things which could happen with that. And to get schools to do transformations, see that things are not scary, and look at Mendelian seggregations. And I think that is the way we need to go.

Peter Kareiva
The world largest environmental group is the Nature Conservancy, it has billions of dollars of assets. It employs 200 Ph.Ds., it has its own postdoc program, it funds research, that's an NGO. That is a major environmental NGO. So we need to have a differentiated view about NGO's.

Klaus Ammann
Yes, I agree, since myself I am a member of IUCN, a global player in Conservation, with thousands of specialists organized in hundreds of committees doing a fine job.

If we want to have a deeper impact on the decision process, we must also involve those who are against genetic engineering. Is that the correct interpretation?

Bernadette Scherer
It is really important that both sides work together. The point which is important for most of the NGO's in Switzerland at the moment has been discussed for about 10 minutes. This is the pollination problem we have with

all this organic farming and for us it is an absolutely major problem. Do we have organic farming in 10 or 15 years if we have transgenic crops around?

Klaus Ammann
Let's go on to conversation about decision procedures – What can we propose from here, from this audience, as new steps? Is it these discursive tactics in the future, and is the strategy to open the dialogue with NGO's?

John Adams
The main topic of a meeting, arranged by Scientists for Global Responsibilities, was global warming and there were some eminent scientists there including a Nobel prize winner. And he claimed that there was two very different versions of responsibility. One was responsibility to the truth as a scientist and the other was responsibility as a citizen. The scientists were attending the meeting, because they thought that the evidence was far from absolutely conclusive on global warming, but on balance, there was something to worry about and they thought as scientists having come to a conclusion that they should seek to influence policy decisions in a direction which would cut down on the emission of CO_2 and so on. Now, once the responsible scientist switches into campaign mode, it becomes extremely difficult.

Klaus Ammann
Scientific knowledge is certainly just one part of the game. When I say code of conduct of course this involves both parties. From what I see is that there are no regulations whatsoever on the side of the behavior of the NGO's. Just absolute liberty, no code of conduct, nothing, and on the other side there are still a lot of people among the biotech companies who think the same. They can do the same. And there, I think, we can make a step forward to talk about certain rules in the dialogue and one of the rules should be respect for the different types of knowledge of the other party, for instance.

Les Levidow
If we look at the papers inside the program, we see representatives of biotech industry, regulators, risk assessment researchers on issues of scientific uncertainties. But there is only one paper presenting views of NGO's, actually describing views of NGO's, and Piet does not represent an NGO as such. So the conference is not intended to discuss the question we now find ourselves discussing. And I have to say I find it offensive to hear such an important political issue be discussed in such a superficial way. I would be rather interested in discussing a code of conduct for regulators and scientists present here.

François Pythoud
If we go as deep into decision making processes as Les is proposing, then we end up by mixing two separate processes: The scientific one on safety and risk assessment and the political questions about what should be the policy of agriculture in the future: We will end up in trouble.

Andreas Seiter
A code of conduct is in my view most probably an urealistic goal, since we need a polarized discussion, where all parties have a chance to express their views, and the public will then make up its own opinion.

Session 7: Methodological lacunas

Methodological lacunas: the need for new research and methods in risk assessment

Gösta Kjellsson

The National Environmental Research Institute, Department of Terrestrial Ecology, Vejlsøvej 25, DK-8600 Silkeborg, Denmark

Summary. The development of biotechnological techniques will allow the industry to produce transgenic plants with altered traits that may affect the chance of invasion and alter organism-interactions in natural communities. In order to assess the hazards from these new transgenes, information on a wide range of subjects is required. It is essential that the most relevant research data are obtained quickly by the use of the most effective research methods available.

Two books with compilations of current test methods useful in risk assessment of transgenic plants were published by Birkhäuser in 1994 and 1997. A total of 161 methods were included covering subjects relating to: invasion, competition, establishment and ecosystem effects [14], and pollination, gene transfer, hybridisation and population impacts [15]. The coverage of different subjects is reviewed, fields are indicated where further research is needed, and the need for new methods is mentioned.

Analysis of recent developments in research literature and trends in releases of transgenic plants in the EU and in USA indicate some major changes we can expect to happen. The importance of resistance to pathogens will increase, the use for production of metabolites, pharmaceuticals, etc. will also increase, and plants with increased tolerance to environmental stress will be produced. The consequences of these aspects are discussed in the context of need of specific information for risk assessment.

Information obtained from literature, field trials, laboratory and greenhouse tests needs to be interpreted in a structured and well defined manner. Hence hierarchical decision procedures and other approaches are needed to meet the demands on "safety" from the public, governmental authorities and international organisations.

New problems arise when transgenic plants are accepted for large-scale commercial releases. This will require that effective monitoring programmes are established for detecting transgene dispersal and unforeseen changes to ecosystems.

Introduction and background

There are many different types of concerns we may face by the use of biotechnology and the introduction of transgenic plants in modern agriculture:

(1) They may pose a risk to human and animal health which requires assessment of the food and feed quality of the new products [19].

(2) There can be social and economic aspects of dependency of certain types of crops over large areas or monopolisation of products by certain companies.

(3) There are ethical aspects which have to be considered. You may object to radical changes in the genome or find that the integrity of natural species are overruled – although manipulation of species has been done for centuries in traditional plant breeding. We have to take these aspects seriously, if public acceptance of the technology is to be gained [5, 8].

(4) Finally, and that is our main object here, there are ecological concerns of risk to the environment. Will there be negative effects to natural habitats if transgenic organisms or genes spread and establish? Many of the concerns and problems expected are similar to those experienced from the introduction of new crops or invasive species [27]. This will, however, depend entirely on the particular transgene insert and the phenotypic effects on the plant. Consequently, our main environmental concern should not be focused on the technique as such, but on the traits and genes that are inserted and on their potential effects to the environment [25].

Now, what are the major potential ecological hazards to the environment from the escape of transgenic plants, the worst things that could happen? The main issues that are commonly quoted [2–4] are:

- Loss of genetic diversity in species - through gene-flow, hybridisation and selection.
- Loss of biodiversity – Species may become extinct by competition or from loss of organism interactions.
- Harm to non-target species – Other groups of organisms (e.g. pollinators and herbivores) may become negatively affected directly by the transgenic plant or if plant species are lost by competition. A change in plant composition in natural habitats may again lead to:
- Changes in primary production, nutrient cycles and geochemical processes, which in the worst case may ultimately lead to:
- Increased soil erosion and other types of soil degradation.

The major international organisations that are contributing to the development of strategies for risk assessment and collecting information on transgenic plants include the OECD, United Nations and the European Union. The OECD contributes to harmonisation of safety measures, workshops, publications on regulatory survey and consensus documents (S. Barber, this volume, [17, 18]). Under the United Nations, UNIDO has established a Biosafety Information Network and Advisory Service, BINAS. United Nations Environmental Program, UNEP, has provided guidelines for biosafety [26]; and global information is gathered in the Information Resource for the Release of Organisms into the Environment, IRRO. EU is central to the regulatory work in Europe with the directives 90/220/EEC and 94/15/EC, the simplified annex for plants, which stipulate the information required for releases of genetically modified plants. These directives are currently under revision. The list of terms now required by the EU is quite good, but exactly what specific information is needed and the type of data is required in each case, is not stipulated.

Need for new research and methods

Exactly what kind of research and methods are needed in the near future? To answer these questions we first look at current lacunas in our knowledge and then try to analyse future needs based on trends in development of biotechnology.

A compilation of current test methods useful in risk assessment of transgenic plants is available in two volumes from Birkhäuser Verlag [14, 15]. In total, 161 methods, some covering a

Table 1. Need for research and new methods in risk assessment of transgenic plants[1]
Subjects covered mainly relate to processes of plant life-cycle, plant competition, population dynamics and ecosystem effects. GMP is used as an acronym for "genetically modified plant" or "transgenic plant".

Suggestion	Purpose
Plant growth and competition	
Need for methods which are less time consuming	Estimate seedling fate in relation to density
New methods for analysis of competitive experimental designs	Generalisation of results
Field methods to study allelopathy need to be developed	Integration of laboratory and field research
Seedling persistence is an overlooked strategy	Survival under stress
Pollination	
Need for information on interactions between GMP and local pollinators	Determines reproductive success of invasive species
Need for research in the role of small, unspecialized pollinators	Influences the range of gene transfer
More data linking pollinator activity and actual gene flow needed	Monitoring gene flow between GMP and are wild species
Information on range and viability of long-distance dispersed pollen is needed	Monitoring long-distance dispersal
Gene transfer and hybridisation	
Lack of data of relative importance of seeds compared to pollen for gene-flow	Estimation of transgene dispersal
Need for additional markers and detailed genetic map to study hybridisation	Monitoring of gene transfer
Need for information and methods to study pleiotropy	Expression of transgenes after hybridisation
Seed dispersal, survival and germination	
Influence of human activity on seed dispersal	Monitoring seed dispersal and invasion
Density of rare seeds in soil difficult to detect	Monitoring seed survival in soil
Artificial seed decay correlated to seed survival in the field	Estimates of survival in seed bank
Need for easy and reliable seed labelling methods	Monitoring seed dispersal distances
Population dynamics and genetics	
Studies linking genetic diversity to variation in plant fitness are needed	Information on the adaptive value of diversity
Need for data on the migration rate dependency of the size of the receiver population	Modelling invasion of GMP into new areas
Knowledge on identity and action of genes causing genetic load is limited	Genetic load may set constraints on invasive success of a GMP
Need for research which integrates genetic and ecological data	Information on fitness of GMP in local environments
Ecosystem effects	
Need to determine habitat dependent success of invasion of crop species	Estimates of invasion probability of GMP
Need for research in palatability of crop species to herbivores compared to wild plant species	Indicates success of invasion and survival
Confined methods concerning effects of perturbations to vegetation are needed	Trials of GMP performance before release

[1]Based on author suggestions in method books [14, 15]

range of subjects, are described with nearly 1700 references providing further information on each method. The subjects that are included in the books are primarily relevant to processes relating to gene dispersal and the life-cycle and population dynamics of the transgenic plant - factors which determine the success of an invasion. For specific subjects, we find that analysis of plant growth and plant competition are well covered with 33 methods. For the study of pollination, 34 methods are available, while gene transfer and hybridisation are covered by 43 methods. A total of 35 methods concern seed dispersal, seed viability (i.e. seed bank) and seed germination. Population dynamics and population genetics are especially well covered subjects with 56 methods available. However, concerning direct ecosystem effects, only 11 methods are available in volume 1. Furthermore, the books do not include methods on toxic effects on organisms or non-target effects caused by food webs, etc., and plant-organism interactions are only represented by plant-pollinator interactions.

In the method books, the authors made several suggestions of specific fields where further research is needed or where there is a need of new methods for risk assessment (Tab. 1). Some of the more general conclusions will be commented on here. For test purposes, there is a need to integrate and validate laboratory research and methods concerning allelopathy and seed decay (survival in seed bank) with data from the field. There is a lack of knowledge on the influence of the type of pollinator and pollinator activity for gene flow and establishment of transgenic plants. Data on the relative importance of seeds compared to pollen for gene flow is needed in modelling. For monitoring of seed dispersal distances and seed survival, methods are needed for labelling and detection of seeds with low density in the soil. The influence of human activities, from cars and agricultural vehicles, etc., on dispersal of introduced species has been largely unnoticed by experimental research, although man probably is a major cause for the dissemination of introduced species. There is a need for research in the function of transgenes which affects the success of invasion (pleitropic effects and genetic load). Research data for modelling is needed on the migration rate dependency of the size of the receiver population. Finally, the integration of genetic and ecological research (e.g. genotype-environment interactions) is needed for conclusions on the fitness of transgenes in local environments.

Need for new research based on recent trends in field releases and biotech research

Current research and development of methods for ecological risk assessment must focus on the specific inserted trait and the particular types of hazards they may cause to the environment. The main traits that are currently inserted by the new biotechnology include: Herbicide tolerance in transgenic plants primarily affects problems of herbicide use in the agricultural ecosystem and adjacent areas on farming practice (e.g. resistance management) and toxic effects on target and nontarget organisms. These problems are only relevant to natural habitats if herbicide use is expected. Pest and pathogen resistance require our fullest attention as these traits directly involve other organism groups, and may also cause indirect effects on non-target organisms (A. Raybould et al., this volume, Fuchs and Gonsalves, this volume). The Bt resistant Maize and the heated debate in EU, is a good example (see also Hilbeck and Bigler, this volume. Altered plant characteristics, such as altered height, flower colour, male-sterility, etc., may directly or indirectly affect the ability for growth, reproduction and influence competitive interactions and invasiveness. Changed metabolic content in the plant (e.g. changed fatty acids, aminoacids, pharmaceuticals, etc.) may cause toxic effects on pollinators, herbivores and soil organisms. Stress tolerant plants (e.g. to drought, salt and cold) may become a very critical group, as they potentially could invade natural habitats along seashores, in northern-temperate or alpine climate, etc., and possibly affect biodiversity and soil processes in the environment. Special problems for humans and livestock health concern the use of antibiotic resistance markers (e.g. Kanamycin); but these will eventually be phased out by the industry.

What type of transgenic plants are currently being released in field trials, and what can we expect in the future?

To answers this question, we compare the percentage releases of different traits in the EU from the period 1991–95 and in 1996 to the data published by the US Department of Agriculture, USDA, for 1987–97 and to the published research literature from the period 1994–95 (Fig. 1).

The development in field releases in EU in 1996 show some minor changes from 1991–95. The percentage of trials with herbicide tolerance and pathogen resistance have both declined, while trials involving new metabolites has slightly increased. The percentage of trials with changed plant characters has significantly decreased. This can largely be explained by the male-sterile rape from PGS with many field releases in the EU in 1991–95.

The data from the USA could indicate what may happen in the EU in years to come. The percentage field releases trials with herbicide resistant crops is much lower in the US than in Europe, only some 29%. Both field trials with metabolic products and insect resistance have greater percentages than in Europe, 21% and 24%, respectively. Pathogen resistance crops are 14% of the trials (9.8% viral resistance and 4.2% fungal resistance) which is similar to the EU. The difference in percentage for changed plant characteristics is partly caused by overlap with the category "Others".

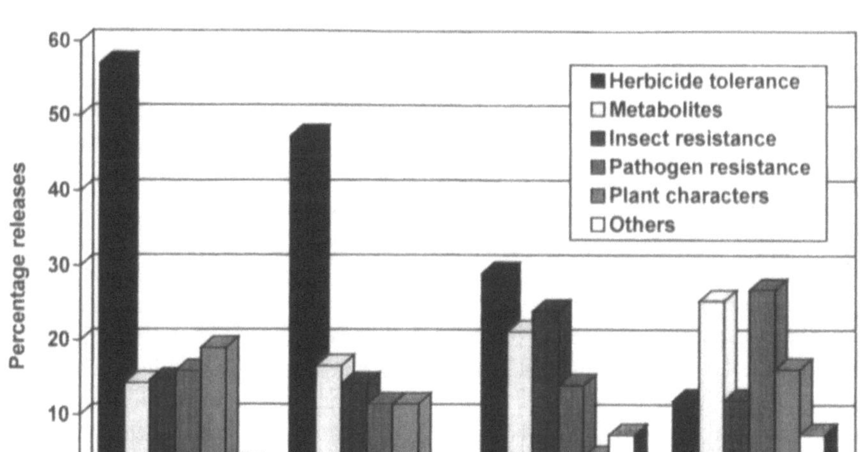

Figure 1. The percentage field releases of transgenic plants with different traits in EU and in USA compared to the corresponding percentages of research literature. Based on information from the EU, DG XI, USDA and a literature search. The percentages for the EU add up to more than 100 because a case may involve more than one trait.

The published research literature can be used as an indicator of what to expect in field trials at a later stage. There are limitations on this approach. The literature search may be biased, and the number of field releases will ultimately depend on what is commercially desirable. The results compared to the EU data showed a decrease of herbicide tolerance to approx. 10% of the published research. Insect resistance also amounts to about 10% of research literature. This comparatively low value may reflect that one technology, the Bt toxins, has been very successful and further research seems less urgent. Pathogen resistance (specifically virus resistance) has increased further to 27% of the references and so has the percentage of studies with metabolites and chemical products in transgenic plant to 25% of references. Others, which include research in stress tolerant plants is becoming more important with 7.5% of the literature references.

To conclude, the survey indicates a trend of decreasing use of herbicide resistant crops in field trials but increasing use of transgenic crops with pathogen resistance (especially virus resistance) and transgenic crops for production of metabolites and chemical compounds. So this is what we can expect to face in future years. How should we act on this information? What should we do?

Need for structured procedures in risk assessment

There is a need to apply more structure in the assessment procedure of transgenic plants than has been done up to now. There is also a need for effective regulation that meet the demands of the industry for clearly stated requirements for information and a limited time scale for regulatory decisions. Criteria for data requirements involving specific test methods are especially needed, which make stricter delimitations compared to present standards (i.e. the EU Directive 90/220, see also [9]). This is required to meet ongoing disputes between countries on which effects to include and how to delimit these in risk assessment procedures, etc. [16, L. Levidow, this volume]. Suggestions on how this can be done in a hierarchical manner, partly based on the experience from risk assessment in ecotoxicology, have been published [24, 23, 13]. The procedures generally involve a number of different tiers or levels of increasing complexity and cost of data and increasing scale of release (small-scale to large-scale field releases). At each tier information and data are collected, tests are made, the results are analysed, and decisions are taken on whether to accept release, reject release or that further data are needed for assessment.

A flow diagram, which suggests how the different types of effects and hazards in risk assessment are interacting, is shown in Figure 2. Not all subjects are relevant to all transgenes or plant species. The diagram may be used to indicate which types of subjects and what kind of tests needs to be done for specific traits and plants. There are some general subjects which directly affect the changes for establishment of the transgenic plant: changes in growth rate, competitive ability and reproduction (i.e. flowering, seed production, seed survival, etc.). Effects of an established population of the transgenic plant on other plant species, vegetation and secondary effects on soil properties, must be considered in tests and monitoring. For specific traits, the issues of non-target effects on pollinators, herbivores, wildlife and soil fauna including secondary food-web effects, must be evaluated. Finally, the issues of gene dispersal and hybridisation, which make further assessment necessary [see 6, 21], has been added to the flow diagram. As an example, for a plant made resistant to insects by Bt-toxins, effects to vegetation

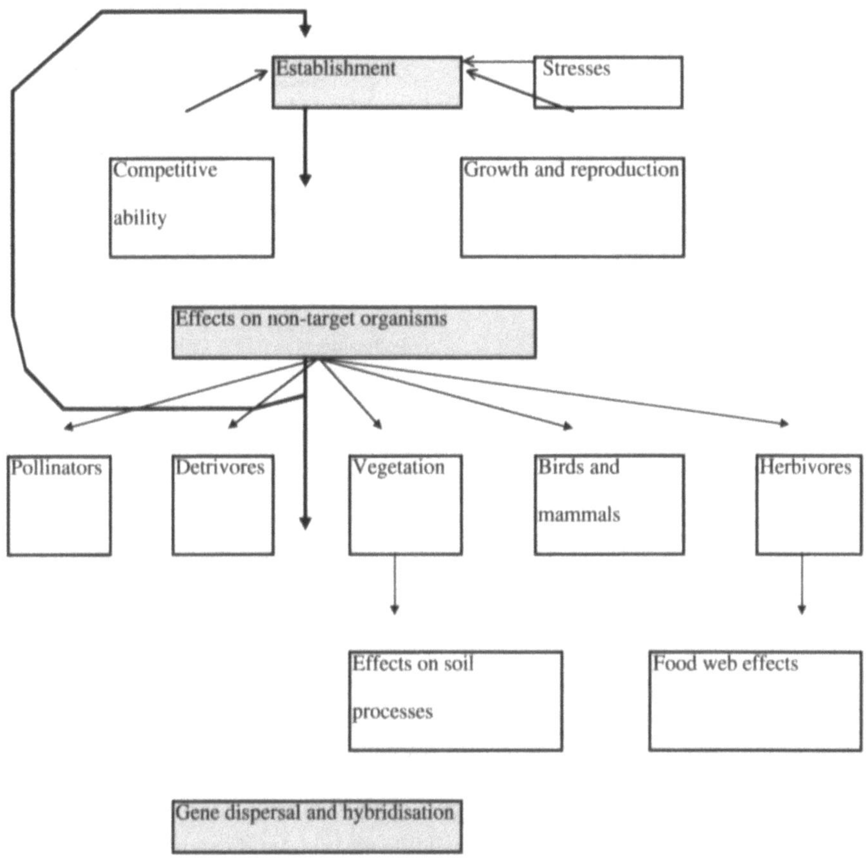

Figure 2. Schematic presentation of main subjects and interactions which should be considered in the environmental risk assessment of a genetically modified plant.

and secondary effects to soil processes would be of little relevance, while effects to other organism groups mentioned above will be important to assess.

It is suggested that an assessment procedure for a transgenic plant should include two types of tests: one package of basic tests, which should cover the general subjects and in principle should be applied to each new case, and one or more packages of tests specific for the inserted trait. All tests should be relatively simple to perform, standardised and replicable. Tests should be done on both the transgenic and the unmodified comparable plant. It is also important to define quantitative or qualitative acceptance criteria, which could be based on decisions made from consensus meetings in international fora of specialist research scientists.

From the basic tests (Tab. 2), data on possible changes of important aspects of plant growth, reproduction and competitive ability are obtained (Fig. 2). Most of these issues are rather fundamental for assessment of survival and establishment in new habitats, and are fully covered in the method books [14, 15]. The tests for the inserted trait should concern the specific type of hazards that may be expected from these traits. It could be important to include a factor of envi-

Table 2. Types of tests for risk assessment of transgenic plants
Suggested basic tests which apply to all cases and an example of a package of toxicity tests for a specific trait, bioproduction.

Basic tests		Toxicity tests	
Subject	Type of test/ data	Possible hazard to:	Type of test/ data
Plant growth	Growth-rate test Total plant biomass	Human health and live stock health	Toxicity and allergy tests (rat-test and skin test)
Plant reproduction (life-cycle)	Seed germination Plant survival Flowering	Wild life	Feeding studies with birds and mammals
	Seed production	Pollinators	Honeybee test
	Seed survival (viability)	Soil organisms	Tests with earthworms and springtails
Competitive interactions	Two-species test	Environmental fate	Degradation tests

ronmental stress when tests are made (e.g. drought, nutrient deficiency or competition). Under optimal growth conditions (normal field trials) critical differences in performance between the modified and unmodified plant may not become apparent. Consequently, these type of results will not indicate what would happen in natural habitats. The type of trial conditions (greenhouse trial or field release) will require that the relevant safety precautions are taken [7].

If we look at a particular case, bioproduction of metabolites, such as proteins, oils, vaccines, etc., besides the basic tests, one package of tests must focus on the possible toxic effects of these compounds (Tab. 2). An array of test methods already exists from the fields of ecotoxicology and food regulation. Aspects on human health include acute toxicity, allergic reactions, cancerogenic effects, etc. Tests of effects on wildlife are relevant if exposure is likely. Possible toxic effects of pollen on pollinators can be tested on honeybee larvae or other insects. Dead plant material may enter the soil environment and cause toxic effects to soil organisms. Environmental fate of toxic compounds can be revealed by tests of degradation in soil under different conditions (soil type, anaerobic).

I will try to pin down what is important in the analysis of effects in risk assessment procedures. First of all we need to define the problem and translate it into the types of measures that give the necessary information. Secondly, relevant test organisms have to be chosen – this should be obvious in most cases of direct effects – or standard test organisms for a certain types of impacts (e.g. resistance) has to be defined. The level of analysis has to be decided – most often individual organisms, populations or ecosystem, which of course is much more difficult to include in tests. An adequate test system (laboratory, greenhouse, field) must be decided on, including the required safety precautions (i.e. confinement, etc.). Relevant test methods for the problem are chosen with the object of testable quantitative results. Pre-trials may be necessary in order to fulfil statistical considerations concerning variation of data (e.g. power-tests). Analysis of results should ideally produce clear-cut answers that facilitate decision-making. Modelling can at least to some extent be used to extrapolate to new situations and a sensitivity analysis will identify key-parameters or forgotten aspects (G. Giddings, this volume). As shown above, this should be done within a structural framework in order to get consistent data from case-to-case.

Large-scale field releases, commercialisation and monitoring

In small scale field releases, the escape of transgenes can normally be prevented by adequate safety precautions (trap plants, exclusion of seed set, large isolation distances, etc.). Furthermore, methods and information on risk of hybridisation and probabilistic models for gene flow and introgression are available for many crops [10, 1, C. Damgaard, this volume, R. Rufener Al Mazyad, this volume, J. Tufto, this volume], although gene flow rates can be highly variable [T. Klinger and N.C. Ellstrand, this volume]. However, in large scale commercial release the transgenes will eventually escape, and even when carefully done, risk assessment cannot be expected to catch all cases of possible adverse effects [11, 12]. Consequently, monitoring is needed to detect unforeseen changes to habitats and other organisms or sudden expansion of the transgenes [20, J.B. Sweet, this volume]. This is also essential for early detection and management to minimise adverse effects to the environment. In this process, definite criteria for long-term impacts to the environment will have to be defined although this may sometimes be difficult [see P. Dale, this volume]. Especially for long-lived perennials, such as trees, the time-scale involved makes quantitative risk assessment difficult [B.Å. Tømmerås, this volume]. Suitable methods for easy detection of spread of transgenes (e.g. by morphological markers such as anthocyanins) and at a low cost may be crucial in certain situations and regions [see V. Simonsen, this volume]. The fact that major impacts of invasions by exotic species have been difficult to detect initially [Marvier et al., this volume] must not circumvent the efforts to develop monitoring methods targeted at determining the effects of transgenic plants. Besides the issue of determining long-term effects, the variation of climatic conditions between years and habitat variability become important issues in monitoring transgene invasion and ecosystem effects [22].

On a larger scale, regional aspects of risk assessment of transgenic plants become important. Differences in climate, topography, soil properties and vegetation between, for example Southern Europe and mountainous Central Europe are surely great. These aspects have influence on gene dispersal rates and the type of hazards which may be expected. Also, the use of a single crop or variety of crop over a large area may affect rate of escape and hybridisation, besides the direct consequences to the farmland. In areas where important crops have their centres of origin, special precautions must be taken to preserve the original gene pools.

To summarise, what we now need to focus on in risk assessment is:

- Specific test methods for each type of inserted traits and the different types of effects.
- Structured risk assessment procedures.
- Monitoring programmes to detect transgene escape and environmental effects when transgenic crops are marketed and used in large-scale releases.
- Internationally accepted procedures and requirements by EU, OECD and other organisations.

However, it is essential that the measures taken get public acceptance at all levels, through debate and a high level of qualified information.

References

1. Ammann K, Jacot Y, Rufener Al Mazyad P (1996) Field release of transgenic crops in Switzerland – An ecological risk assessment of vertical gene flow. *In*: E Schulte, O Käppeli (eds): *Gentechnisch veränderte krankheits- und schädlings-resistente Nutzpflanzen.* Schwerpunktprogramm Biotechnologie des Schweizer-

ischen Nationalfonds, Bern, 101–157
2. CDPE (1992) *Ecological impact of genetically modified organisms. A survey of literature, guidelines and legislation.* Steering committee for the conservation and management of the environment and natural habitats. Council of Europe, Strasbourg
3. Colwell RK (1994) Potential ecological and evolutionary problems of introducing transgenic crops into the environment. *In*: AF Krattiger, A Rosemarin (eds): *Biosafety for sustainable agriculture: sharing biotechnology regulatory experiences of the western hemisphere.* ISAAA and SEI, Ithaca, USA, 33–46
4. Crawley MJ (1995) Long-term ecological impacts of the release of genetically modified organisms. *In*: *CEP Pan-European conference on the potential long-term ecological impact of genetically modified organisms.* Proceedings Strasbourg, 24–26 November 1993. Council of Europe Press, Strasbourg, 43–66
5. Daele W van den Pühler A, Sukopp H (1997) Part III: normative evaluations – ethics, law and politics. *In*: Daele, W. van den, Pühler, A., Sukopp, H. *Transgenic herbicide-resistant crops. A participatory technology assessment.* Wissenschaftszentrum Berlin für Sozialforschung, Berlin, 59–99
6. Dale PJ (1994) The impacts of hybrids between genetically modified crop plants and their related species: general considerations. *Mol Ecol* 3: 31–36
7. Dale PJ, Kinderlerer J (1995) Safety in the contained use and the environmental release of transgenic crop plants. *In*: Tzotzos, G.T. (ed.): *Genetically modified organisms – a guide to biosafety.* CAB International, Wallingford, 36–63
8. EC (1997) European opinions on modern biotechnology. *Eurobarometer 46.1.* European Commission, Directorate General XII, Brussels
9. Gaugitsch H, Torgersen H (1995) Streamlining regulations, keeping high safety standards: Revised criteria for the assessment of releases of genetically modified organisms (GMOs) into the environment. *Ambio* 24: 47–50
10. Gliddon C (1995) The potential role of monitoring in risk assessment and in fundamental ecological research. *In*: *CEP Pan-European conference on the potential long-term ecological impact of genetically modified organisms.* Proceedings Strasbourg, 24–26 November 1993. Council of Europe Press, Strasbourg, 253–263
11. Kareiva P, Morris W, Jacobi CM (1994) Studying and managing the risk of cross-fertilization between transgenic crops and wild relatives. *Mol Ecol* 3: 15–21
12. Kareiva P, Parker IM, Pascual M (1996) How useful are experiments and models in predicting the invasiveness of genetically engineered organisms? *Ecology* 77: 1670–1675
13. Kjellsson G (1997) Principles and procedures for ecological risk assessment of trans-genic plants. *In*: Kjellsson, G., Simonsen, V., Ammann, K. (eds): *Methods for risk assessment of transgenic plants. II. Pollination, gene-transfer and population impacts.* Birkhäuser, Basel, 221–236
14. Kjellsson G, Simonsen V (1994) *Methods for risk assessment of transgenic plants. I. Competition, establishment and ecosystem effects.* Birkhäuser, Basel
15. Kjellsson G, Simonsen V, Ammann K, eds (1997) *Methods for risk assessment of transgenic plants. II. Pollination, gene-transfer and population impacts.* Birkhäuser, Basel
16. Levidow L, Carr S, Wield D (1997) European biotechnology regulation: contested boundaries of environmental risk. *BioSafety* (Online Journal) vol. 3, paper 1
17. OECD (1992A) *Report on the OECD workshop on the monitoring of organisms introduced into the environment.* OECD, Paris
18. OECD (1992B) *Safety considerations for biotechnology.* OECD, Paris
19. OECD (1993) *Safety evaluation of foods produced by modern biotechnology: concepts and principles.* OECD, Paris
20. Parker IM, Bartsch (1996) Recent advances in ecological biosafety research on the risks of transgenic plants: A trans-continental perspective. *In*: Tomiuk, J., Wöhrmann, K., Sentker, A., (eds): *Transgenic organisms – Biological and social implications.* Birkhäuser, Basel, 147–161
21. Raybould AF, Gray AJ (1994) Will hybrids of genetically modified crops invade natural communities? *Trends Ecol Evol* 9: 85–89
22. Regal PJ (1994) Scientific principles for ecologically based risk assessment of transgenic organisms. *Mol Ecol* 3: 5–13
23. Rissler J, Melon M (1993) *Perils Amidst the Promise. Ecological Risks of Transgenic Crops in a Global Market.* Union of Concerned Scientists, Cambridge, MA
24. Suter GW (1993) *Ecological risk assessment.* Lewis Publishers, Boca Raton
25. Tiedje JM, Colwell RK, Grossman YL, Hodson RE, Lenski RE, Mack RN, Regal PJ (1989) The planned introduction of genetically engineered organisms: Ecological considerations and recommendations. *Ecology* 70: 298–315
26. UNEP (1996) International technical guidelines for safety in biotechnology. *United Nations Environmental Programme*, Nairobi, 31
27. Williamson M (1993) Invaders, weeds and the risk from genetically manipulated organisms. *Experientia* 49: 219–224

Methods for Risk Assessment of Transgenic Plants
III. Ecologıcal risks and propspects of transgenic plants
K. Ammann, Y. Jacot, V. Simonsen and G. Kjellsson (eds)

Discussion session 7: Methodological lacunas

Jörg Landsmann
The extensive list of potential hazards Gösta Kjellsson gave us in his contribution suggested that they were specific to GMO's. However, I think this is a random selection of hazards from everyday agriculture. Some of these hazards actually increase with genetically modified plants, but others may even decline.

Gösta Kjellsson
Well, I fully agree with you. For one thing you could argue there is no difference between the problems we face with the transgenic plants and the ordinary crops. And you have to be very specific in assessing these problems.

Andreas Seiter
I have a similar observation. It is very difficult for me to distinguish between risks which are specific for transgenic plants and risks which are kind of general agricultural risks. And I have a proposal to make, which I would like to have briefly discussed by the speakers this morning. In medicine sometimes you compare two different situations which are comparable in certain parameters. So why not compare a specific new introduced feature of a transgenic plant with a similar feature in a traditionally bred plant, like insect resistance introduced by transgene compared with the ecological impact of traditional insect resistance. We should realise that some of those risks have already existed for 50 years.

Gösta Kjellsson
Yes, and this is just short response. It is a wise idea to have risk assessment for conventional crops also. It is also necessary to consider the differences of the trait inserted, the gene and the toxic effect as in Bt.

Klaus Ammann
You remember my statement from yesterday. Genetic engineering is a wonderful tool, actually a marker technology, in order to detect and to monitor what goes on also in classic agriculture.

Jan Carel Zadok
Yesterday, I was very critical about monitoring – that is based on my experience as a plant pathologist. And I emphasise the point that you have to be very precise in research questions which you can test. Then the answer can be tested statistically. The method must be reproducible and transferable. Now here Dr. Kareiva has indicated an approach which would even suite the critical plant biologist. I'm very anxious to see his book in print [PLFiedler PM Kareiva (eds) (1998) *Conservation Biology: For the Coming Decade*. Chapman and Hall, New York]

Tom Nickson
Returning to Dr. Ammann's talk, some of which dealt with a step-by-step approach.

When I listen to the many speakers, monitoring seems to develop to quite a daunting task for science, industry and regulators, without any obvious timelines, without any obvious hazards denominated. We certainly need a good definition of what risk really is. Is it risk to human health, it is risk to animal health? Unfortunately the farmers are not represented here. Because, the fact of the matter is, they rotate crops, and these risk assessments then have to consider this. We also have to consider risk balance, we really want to do what is appropriate and what is right for the environment. I have not seen a good step-by-step approach here. Now under monitoring specifically, I think that there are some really good ideas there. But I caution about some of the terminology and the endpoints again, which creates the base problems we have. Environmental disasters would be qualified as when you have transgenes massively invading ecosystems and having a considerable negative impact at the same time. These things will all take time to truly understand and understand their importance. It is an interesting idea to start using the EPA's paradigm for monitoring, but the EPA has real advantage over the ecology in the sense that they have very well defined risks and they have LD 50. They have tests regarding human health done by specific toxins, but it is hard to get a handle on what these toxins really will do in the environment. I am trying to think of what we can really do on a step-by-step basis to move this forward.

Klaus Ammann
This is what we need to do, because we want some outcome and we want some steps to be proposed. We need to learn, just as Peter Kareiva said, from the dynamics which exist already. He incidentally also suffers from the bias in academic biology policy, where observational strategies are dismissed as being non scientific. I think there we need to learn from each other.

Peter Kareiva
I want respond to the last remark. Most of my work is with endangered species. I adopt that principle which is, you find out at the beginning what is practical and then you go from there. So for situations you find out what data you can really get and what you are monitoring and what risk assessment has been done there. It is important to have a quick starting point. First you agree on what is feasible and then you say how you can guess or implement this: It has nothing to do with special coverage of a weed or a gene or an environmental disaster you have to envisage. If there is going to be an environmental disaster, you would have to have extensive special coverage anyway.

Christian Damgaard
This also applies to a catastrophe. You had the analogy of an epidemic situation and I think that is an exaggeration of the problem. It is like comparing a competition model with an epidemiology model, which I think is not appropriate.

Jim White
We have talked often about what kind of controls and what comparisons you use. I have read a paper from Peter Kareiva and other colleagues, Allison Snow on commercialisation of transgenic plants, potential ecological risk. Well, evolutionary effects of engineered crops exacerbate weed and pest problems. Now I want to quote a paragraph: It says, that in general, "there are few examples of weeds benefiting from specific fitness related crop genes. This could be due several factors: the lack of attention, the absence of crop genes that influence fitness advantages to wild relatives – or simply the fact that the impact of beneficial genes is not dramatic."

Klaus Ammann
Since this is a dialogue between science and biotech companies. Is there anything to say about this and regulators, of course?

Tom Nixon
I have not heard anything that I felt was absolutely unreasonable and many things we have not heard before. I think there are some certainties, some opinions of what does not need to be shared in clarifications, that need to be made at a very exact scientific level, but I do not think that I've heard anything that really means that we can anticipate progress in an area. But I want to be really clear at least in terms of the monitoring. It has got to be clearly understood that for any kind of a company dealing with pesticides (a Bt-plant is a pesticide as well as Roundup for us) that have any in all data that we generate or are aware of impacts to the safety of humans and environment, we are obligated by law to report that. And we do report those things, we and everybody else who is in this industry. It is part of our legal obligation and it is also because Monsanto has a council for stewardship that really goes beyond the corporate strategy for marketing. It is their job to make sure that Monsanto is behaving as a steward in any environment and area in which they participate to make sure that we are doing the correct things. So that covers the aspect of imposed reporting which should be considered in any kind of plant for monitoring. I think it is a great idea to have independent people doing data, call it monitoring or call it academic research, that is sometimes supported by the industry and sometimes supported by the science foundations. That is collaboration between industry and academia and that should continue and will be supported. So I would really like to see some good examples of monitoring that fit at the regulatory level. I would like to see really good proposals for extending the science. We do not know everything at the scientific level, and we will never know everything about any product we did not put on the market.

Jan Carel Zadok
I apologise for becoming poetic on a Saturday morning, but as a regulator I want to make a statement that we always have to deal with specific applications. And specific risk analysis and very limited time frames. So we can not go through all the ecology text books and we should stick to the "need to know" so that list may grow.

Session 8: Conclusion, strategies, where do we go from here?

Methods for Risk Assessment of Transgenic Plants
III. Ecological risks and propspects of transgenic plants
K. Ammann, Y. Jacot, V. Simonsen and G. Kjellsson (eds)

Where do we come from, where do we go from here?

Klaus Ammann and Biljana Papazov Ammann

Botanical Garden, University of Bern, Altenbergrain 21, CH – 3013 Bern, Switzerland

How can we move from knowledge to action?

There are of course *tame problems* where you only need a piece of paper, write down the problem and its solution, hand it over to executives and you can be certain of a positive outcome. Sometimes you need more than a piece of paper, but rather a visit paid to local authorities to talk them into accepting an obvious solution. You can even make it a rule that properly describing a problem in all its aspects is often the same as solving it. As a rule, tame problems can be defined in quantitative parameters. If you can present comprehensive scientific data, the solution is nearly automatically defined.

But unfortunately, most essential planning problems are *wicked problems* with complex structures and no obvious causal chains. These problems cannot be determined totally in a quantitative way, there are no solutions existing in the sense of definitive and objective answers.

Unfortunately, wicked problems have been treated mainly in two directions:

- Through formalised methods which are suitable only for the solution of tame problems.
- Often solutions have been found empirically; with trial and error acceptable solutions can be found, gifted planners often develop good intuition taking into account also socio-economic factors. But deplorably, too often the systems approach which works properly for tame problems ends in a fiasco when tackling wicked problems.

Systems approach of the first and second generation

Much hope has been placed in the *systems approach of the first generation*, which certainly had its merits (NASA missions, toll bridges, defence systems etc.). Planning goals were "clearly" defined and all decisions were oriented towards these goals.

In general it can be said that the systems approach of the first generation has been followed by an era of disappointment, since it has not yielded what was expected of it: A number of large and complex projects such as urban renewal, improving the environment, tackling the nutrition problems of mankind etc. can only be considered as failures or partial failures such as the "green revolution".

The main reason is the fact that the classic paradigm of (rational) Science and Technology is not applicable to the problems of open ecological and/or societal systems. It is very important to realise that problems in biotechnology are not solely problems of Science, but also problems of society. This does not mean that risk assessment should not be Science based, on the contrary. It would be a big mistake to assume that the involvement of open structures in ecology and human society would give cheap excuses to deviate from the path of Science, when it comes to questions of safety and regulation.

Professional management tools which are based on a systems approach of the second generation should not be mixed up with *"Future workshops"* with their frequent and inconsiderate use of pin walls when activist groups start their "planning". Rarely those actions have led to sustainable results, too often future-workshops (German *"Zukunftswerkstätten"*) start with a fulminate brain storming and lots of enthusiasm and later on the participants go home to live their normal lives and they tend to forget about their big decisions made earlier. If the workshops would be properly carried through, after Müllert and Jungk, results would be certainly better.

We should also see the differences of *"Collaborative Learning Workshops",* which can be very delightful and thus also successful for reaching decisions, but rarely do such events achieve sustainable results. They exactly lack the process of collaborative decision making. (It is important to avoid a misunderstanding: Decision making is not in its basic structure a democratic process, it is a process where people genuinely involved are participating. To be even more explicit: Partners taking part in the decision-making process should have their own and a genuine interest in the cause, this avoids the danger of manipulation through clever public relations, through populistic and fundamentalistic argumentation.

Consensus conferences and also *citizens conferences* are extremely helpful in cases of conflicts among the public, but here again it is difficult to see that processes criticised will be changed to the better and negative trends are definitively turned around.

Another kind of internal consensus conference is designed by the promoters of the "Syntegrity approach", which brings together corporate people in order to analyse internal dynamics and processes and to discern negative effects.

Despite the fact that there is a lot of effort becoming evident to design new planning and management methods, negative results are predominant and are in fact part of a planning crisis, stemming from the 1970s and which is still continuing today.

What is the "Systems approach of the second generation"?

Still, it is primarily the paradox of rationality which has been severely underestimated in the systems approach of the first generation.

The more questions we ask the more possible answers and *vice versa.* Limitations of technological solutions are always hidden in the open ecological and social systems: Constraints in possible secondary effects in ecology, finances, public perception or changed priorities with time will make the difference. In other words: Decisions taken without considering secondary effects in ecology, finances and public perception will not lead to sustainable solutions.

- In order to tackle wicked problems you need to go through *an extensive process of argumentation*, also called objectification, not to be confusion with an "objective approach" to the problem.

There is rational planning, but there is no way to start to be rational, one should always start a step earlier, since there are important trends and facts which will make straightforward rational thinking and acting in solving wicked problems useless. It is not the theory component, but rather the political component of the knowledge, which determines the vector of the action. This is the *zero step* so important in the publications of Horst Rittel. This is also the basis of the understanding of the term *"Symmetry of Ignorance".*

As an example: The fact that experts can be wrong and farmers know better in certain situations in agriculture because farmers are better observers out in the field.

The knowledge needed in wicked planning problems is not concentrated in a single head. It is absolutely essential to let all partners be involved in the problem solution process, which includes part of the population (mainly farmer organisations and consumer organisations), the Governmental Regulators, the Non-Governmental Organisations, the Life Science Companies and the Scientists. There is no monopoly of knowledge. Having illustrated the difficulties in solving wicked problems, we need a new approach in problem solving, in order to avoid the pitfalls of ignoring bottom-up feedback.

You only can keep to this rule if you are also following another important rule: All partners in the planning process have to avoid hidden agendas, which is certainly eased by respect paid to each other partner. Nobody should be criticised for speaking up about personal interest.

It is obvious in these times of growing difficulties in communicating about biotech products, especially in agriculture, that all partners still have a lot of homework to do.

- The Biotech Companies are populated with people who are convinced about their own products, since they are now precisely concerned about safety standards and regulatory processes. So far so good, but these people live in a world of euphemisms and perfection, they develop with time a lack of understanding criticism from outside.
- The scientists often are naïve enough to stick to factual, instrumental and explanatory knowledge alone. Many miss a very important point as Hannah Arendt put it: One of the most noble tasks of scientists is to make out of facts public opinion.
- The regulators should find ways and means to cope with the growing speed of new developments. One of the main reasons why things in Europe turns sour is the fact that European regulation is way behind that of the United States. On the other hand, this is an excellent occasion to see more clearly the geographical differences in the regulation.
- Some of the NGO's have developed into powerful protest industries and are not interested in a thorough scientific analysis, since this could blur populist argumentation, which they need to keep up in order to get more donors.
- The public is often lost between the two camps and, surprisingly enough, only a minority feels the need for better education, whatever this would mean according to the two camps described above.

An example of argumentative planning method

This process oriented argumentative planning method has been put into effect by a group of *UNESCO summer school* participants of planners, geographers and ecologists on the island of *Scopelos in Greece* where some 16 issues have been discussed in order to improve the ecological situation of the island. Some of the issues were:

- Shall tourism on Scopelos be promoted?
- Which kinds of tourism is/could be on Scopelos?
- What should be the priorities between national and local interests regarding tourism?
- Who are the major decision makers who influence the development of the island?
- How can the environmental awareness of the island's population be improved?

Each issue was tested against a variety of answers which themselves led to a whole chain of arguments for and against answers, and finally, the summary contained some surprising solutions, which were never thought of when the planning process started: Extension of family run hotels, eco-tourism which exposes the tourist to nature trails, scenic views, monastery visits, wine tasting, in-shore-fishing, educational studies of flora and fauna.

But the most surprising solution popped up at the end when the primary agricultural sector was finalised: It was decided to reorganise the *breeding* of the meagre and small *Aegean Cow*, whose genetic resources were nearly lost. This cow has the unique ability of grazing on rocky meadows in a way which allows reforestation, and, surprisingly enough, provides modest amounts of milk for the small farms.

How to solve wicked problems in biotechnology and the environment

What we need in such cases is an action oriented approach. Risk Assessment and Management must be seen as a planning strategy of the second generation in developing a professional framework for *decision making*.

Strategies have to be developed to recognise the consequences of our doing on one hand, and to specify our knowledge on the other hand. This knowledge has to be gained step by step and case-by-case: If we want to clearly distinguish our present state of knowledge for appropriate decisions to be made *not* based on our views and opinions, we need to go through the following steps.

- What is the problem?
- What do we want?
- What are the alternatives?
- How do we compare them?
- How can we reach the solution?

All participants need to keep in mind that there are *various types of planning knowledge* (arranged according to the five questions asked above):

Examples given here are lumped together as simple keyword-illustrations, taken out of their context in real planning examples, they cannot be regarded as an example of a realistic situation, this would be exactly the task of a planning process of the second generation.

- *Factual knowledge* is the knowledge of what actually happens (quantitative data or empirical, observational data).
  Gene flow species by species / region by region / facts about insect resistance in agriculture.
- *Deontic knowledge*, the very important knowledge of what ought to be.
  The knowledge about new crops which enhance agricultural production / new agricultural techniques to avoid erosion / new biological approaches to fight insect pests etc.
- *Explanatory knowledge* explains why things are so or why certain effects will happen. Here already you start to determine the direction of the solution.
  The way Bt proteins are acting on specific pest and beneficial insects / what are the main reasons of unwelcome erosion effects / mechanisms of vertical gene flow / mechanisms of resistance development.
- *Instrumental knowledge* on how to steer certain processes, on how to achieve certain goals, knowledge which needs to be balanced against regulation and safety.
  How to build Bt and other genes into crops and how to stabilise them / how to avoid vertical gene flow / how to avoid unwelcome soil erosion / how to avoid early upcoming pest resistance.

- Conceptual knowledge which would allow to avoid conflicts before they pop up. This is the knowledge about complex situations, taking into account all previous kinds of knowledge and also weighting them against arguments coming from open ecological and societal systems.

  Concepts about transgenic crops compatible to the ideas of a sustainable agriculture.

You need to go through an *extensive process of argumentation*, also called objectification, not to be mixed up with an "objective approach" to the problem. The hopes of this process are:

- to forget less, to raise the right issue
- to look at the planning process as a sequence of events
- to stimulate doubt by raising questions, to avoid short-sighted explicitness
- to control the delegation of judgement: Experts have no absolute power, scientific knowledge is always limited.

There is no scientific planning

Solving practical problems to develop sustainable transgenic crops cannot be dealt with by "scientification of planning". Dealing with wicked problems is always political because of its deontic premises (meaning that you have to involve knowledge of what ought to be). Science only generates factual, instrumental and in the best case, explanatory knowledge.

The planner (here the manager of an action plan) is not primarily an expert, but a *"mid-wife of problem solving"*, a teacher more than a doctor. Moderate optimism and careful, seasoned respectlessness are required, casting doubt is a virtue, not a failing in an action plan manager.

The planning process of wicked problems has to be understood as an *argumentative process*, it should be seen as a venture (or even adventure) within a conspirative framework, where one cannot anticipate all the consequences of plans.

Systems methods of the *second generation* are trying to make this deliberation explicit, to support it and to find means in order to make this process more powerful and to get it under better control. *Planning is an argumentative process.*

Outlook

It is beyond logic and present day knowledge to predict some surprising outcomes in genetic engineering debates designed as above. Still there are some dreams and hints, which should be placed at the end of this contribution:

- *"Precision Biotechnology"* could lead to a better design of crop seeds in future. Precision biotechnology would mean that a bag of seeds contains a great variety of different kinds of seeds related to resistance against many pest insects on one side, but all having a precisely designed genome for the product quality to be sold after harvest.
- *Organic farming* needs in the future to go together with modern breeding methods including genetic engineering. This is in the eyes of the authors an absolute need but also a very difficult thing to achieve, since the transgenic crops of the first generation are either not made for the strategies of organic farming or even worse, they work against such visionary strategies.

Maybe we need some newly designed products which will fit to terms like *Organo Transgenic Crops* and *Organic Precision Biotechnology?*

Methods for Risk Assessment of Transgenic Plants
III. Ecological risks and propspects of transgenic plants
K. Ammann, Y. Jacot, V. Simonsen and G. Kjellsson (eds)

From risk assessment to a more comprehensive technology evaluation – contribution of modern plant breeding to sustainable agriculture

Elisabeth Schulte

Fachstelle für Biosicherheitsforschung und Abschätzung von Technikfolgen des Schwerpunktprogrammes Biotechnologie des Schweizerischen Nationalfonds (BATS), Clarastrasse 13, CH-4058 Basel, Switzerland

Genetic engineering will become a key technology in the future. New breeding methods aided by recombinant DNA technique will provide a significant potential for the realisation and improvement of important agricultural products. Modern biotechnology not only offers unprecedented opportunities for studying and understanding biological principles and processes, but it also makes available a library of powerful tools, which complement the methods of traditional technologies. The first transgenic crop varieties are now entering the marketplace and first significant changes in conventional agricultural production are becoming apparent.

Technological developments are characterised by ambivalence and by wider impacts than the original target purpose, as reflected in the public discussions on potential risks and benefits. Even extensive safety evaluations, carried out in order to gain the approval of the regulatory bodies, cannot comply with safety perceptions excluding any harm for a technology application. Absolute safety does not exist, and the most comprehensive safety evaluations still must acknowledge the possibility of unforeseen effects arising from the natural genomic variability of living organisms or from the complex biological processes within ecosystems. The challenge will be to determine the degree of uncertainty which could be accepted.

The acceptance of a technology is strongly influenced by its benefits, especially its contribution to a desirable development of environment and society. This would suggest a more comprehensive impact analysis comprising ecological, economic and social components in comparison to those of different technical options.

Controversial discussions about the contribution of gene technology to plant breeding are to a great extent discussions made of individual convictions about the *right way in agriculture and food production* (Fig. 1). The various concepts of agriculture range from high-tech agriculture and integrated production to organic farming.

An evaluation of different options demands a *reference* or in other words some idea of *how* future food and agricultural production should proceed. The need to move toward a more sustainable production – one which is environmentally sound, economically viable, and socially just – has become more and more urgent. In that context the evaluation of new technologies, such as modern plant breeding, in terms of the underlying risks and benefits becomes a basis and on top of it a new dimension.

At the 1992 Rio Earth Conference, the international assembly of states agreed to elaborate corresponding concepts for a sustainable development. Because countries and even regions within the same country have different social, economic and environmental contexts, a unique definition of sustainability cannot exist. Sustainability should be regarded in a system-based approach, where the dynamics and inherent complexity of agroecosystems have been taken into account (Fig. 2).

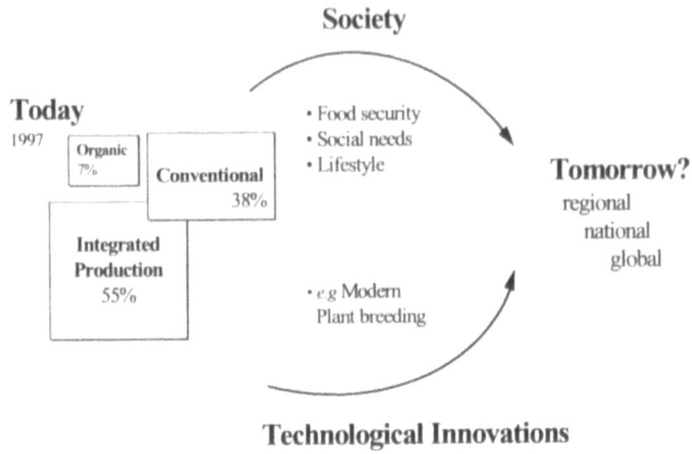

Figure 1. *Driving forces* for tomorrow's agriculture. The distribution of different agricultural production methods inSwitzerland is shown here.

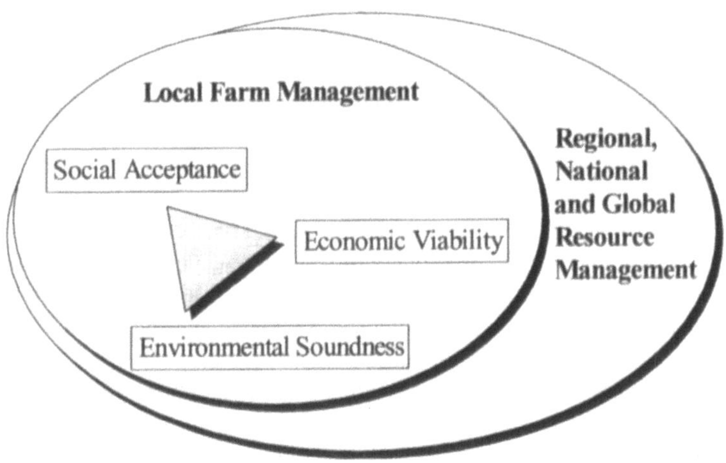

Figure 2. The spatial dimensions of sustainability encompassing social, economic, and environmental components.

The realisation of this concept has to be adapted to suit local needs, without losing sight of the necessity for a regional, and finally a global approach to sustainable resource management. In order to evaluate modern biotechnology for its contribution to sustainable agriculture, detailed criteria and indicators need to be defined. Indicators are useful tools to gain insight regarding the progress made in achieving sustainable development. Not all of the indicators will be applicable in every situation. Countries will choose those relevant to national priorities, goals and targets.

A comprehensive technology assessment of the introduction of transgenic crops to Swiss agriculture was initiated in 1994 (Fig. 3). During a first phase (1994–1997) potential impacts

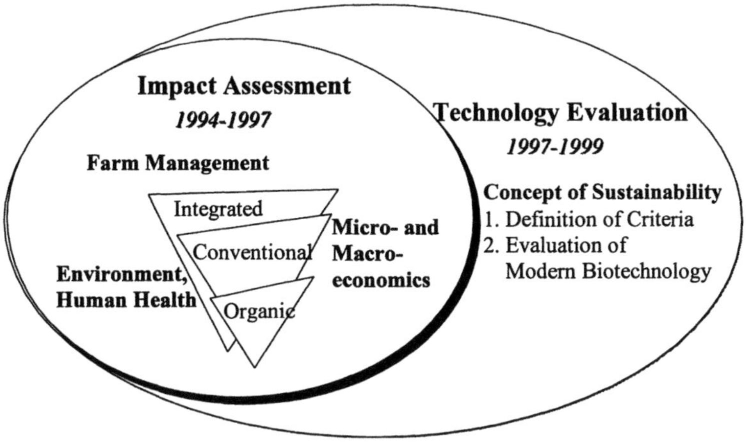

Figure 3. Case study in Switzerland: Technology Assessment on the contribution of modern plant breeding to sustainable agriculture.

on environment, farm management, and micro- and macro-economics were analysed [1]. In the second phase (1997–1999) the contribution of modern plant breeding to sustainable agriculture is currently being evaluated [2]. Participating actively in this project are a multidisciplinary group of scientists, and an accompanying committee consisting of representatives from regulatory bodies and of private and public interest groups.

References

1. Schulte E, Käppeli OGentechnisch veränderte krankheits- und schädlingsresistente Nutzpflanzen Eine Option für die Landwirtschaft? (1994–1997) Schwerpunktprogrammes Biotechnologie des Schweizerischen Nationalfonds Bern
Band 1 (1996), Materialien; 632 Seiten BATS, Basel
Band 2 (1997), Auswertung; 45 Seiten, BATS, Basel
2. Schulte E., Käppeli O. *Nachhaltige Landwirtschaft – Kriterien für Pflanzenzüchtung und Pflanzenproduktion unter besonderer Berücksichtigung des Potentials der modernen Biotechnologie* (1997–1999) Bundesamt für Landwirtschaft, Bern; Schwerpunktprogramm Biotechnologie des Schweizerischen Nationalfonds, Bern; BATS, Basel

Methods for Risk Assessment of Transgenic Plants
III. Ecological risks and propspects of transgenic plants
K. Ammann, Y. Jacot, V. Simonsen and G. Kjellsson (eds)

Transgenic plants and the management of virtual risks

John Adams

University College London, Department of Geography, 26 Bedfordway, WC1H OAP London, UK

Risk management involves *a balancing act* in which the potential *rewards* of a contemplated action are weighed against the potential *costs*. There has been a long-running and sometimes acrimonious debate between "hard" scientists – who treat the rewards and costs as capable of objective measurement – and social scientists – who argue that risk is culturally constructed. Much of this debate evaporates if one distinguishes three categories of risk:

directly perceptible risks: e.g. climbing a tree, riding a bicycle, driving a car,

risks perceptible with the help of science: e.g. cholera and other infectious diseases,

virtual risks – scientists do not know or cannot agree: e.g. low-level radiation, pesticide residues, global warming.

In Figure 1 these categories are represented by three overlapping circles to indicate that the boundaries between them are indistinct, and also to indicate the potential complementarity of approaches to risk management that have previously been seen as adversaries.

Directly perceptible risks are managed instinctively and intuitively. We do not undertake a formal probabilistic risk assessment before we cross the street. Other risks are only perceptible with the help of science. With a microscope, for example, one can see, and measure objectively, the agents responsible for many infectious diseases. To the extent that science illuminates for non-scientists the connections between behaviour and consequence, it shifts risks into the directly perceptible category. But there remain many risks about which scientists cannot agree. Many of the risks associated with genetic modification fall into this category. These risks relate to potential health effects, to the potential loss of control over environmental releases, and to the concentration of power over the processes and products of genetic manipulation.

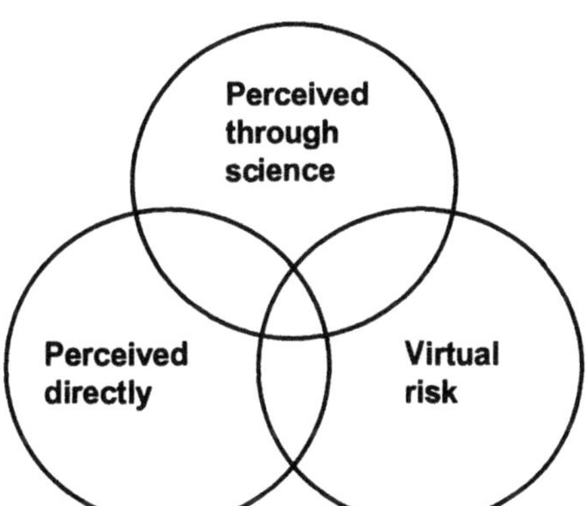

Figure 1. Three types of risk.

Some of the risks associated with food can be assigned to the category of *risk directly perceptible*. Our senses of sight, smell and taste form our first line of defence against food that might make us ill. Putrid food offends all three senses and is rejected. Commonly the rewards are also directly perceptible; eating is one of life's pleasures and we are attracted to foods that look, smell and taste delicious. Hunger and our sense of repleteness also govern, more or less satisfactorily, the quantities we consume.

Science also plays an important role in what we eat. Folk science, in the form of accumulated knowledge about which plants are poisonous, or curative, has assisted direct perception for many millenia. Increasingly the range of direct perceptions is being extended by the printing on packaging of use-by dates and other advice relating to preparation and nutrition. Modern science in the form of knowledge about poisons, vitamins, allergies, metabolism, genetic susceptibilities etc. also guides the regulators of the food chain. But at the same time that science is illuminating and reducing old risks, it is creating new ones. It produces impressive rewards – in the form of nuclear power, new materials, effective pesticides, new crops etc. – but is often accompanied by uncertain, and potentially catastrophic, side-effects.

We do not respond blankly to uncertainty, we impose meaning upon it. Long-running controversies about large scale risks are often long running because they are scientifically unresolved, and unresolvable within the time scale imposed by necessary decisions. The clamorous debates that take place in the presence of uncertainty are characterised not by irrationality, but by *plural rationalities.* Scientific uncertainty liberates people to argue from pre-established beliefs, convictions and biases. The contending parties often argue logically, but from different premises. Figure 2 presents examples of responses characteristic of well-established biases.

- *Individualists* are enterprising "self-made" people, relatively free from control by others, and who strive to exert control over their environment and the people in it. They are pragmatic and optimistic, and tend to be more impressed by the potential rewards of genetic science and technology than by its risks. Nature, according to this perspective, is to be *commanded* for human benefit.
- *Hierarchists* inhabit a world with strong group boundaries and binding prescriptions. Social relationships in this world are hierarchical with everyone knowing his or her place. They are the regulators responsible for containing the risks associated with genetic manipulation. Nature is to be *managed.*
- *Egalitarians* have strong group loyalties but little respect for externally imposed rules, other than those imposed by nature. They are democrats who resent and fear the power of both big business and big government. Nature is to be *obeyed and respected and interfered with as little as possible.* The activities of the large bio-tech industries are resisted as *unnatural* and *disempowering*.
- *Fatalists* have minimal control over their own lives. They belong to no groups responsible for the decisions that rule their lives. The best you can do is *duck if you see something about to hit you.*

As the science becomes less certain, the importance of these biases increases. In brief, we live in an uncertain world, but certain conclusions about the management of genetic risks might, nevertheless, still be ventured: It is important to be clear about the nature of the risk under discussion.

☹ **Fatalist** • *The whole world is powerless to countermand the actions of powerful, profit-driven corporations:* "[GMOs are] being inflicted on unwilling people like myself by Monsanto's unwelcome inclusion of GMOs in the world's food supply... There are no benefits for the consumer by the inclusion of GMOs, only greater profits for Monsanto."  Austin, *The Guardian*, 16 December 1997. Gallows humour is a common fatalist response to perceived powerlessness.	☺ **Hierarchist** *genetically modified organisms constitute a management problem, soluble by science and regulation* "We conduct a full scientific risk evaluation. Once we are satisfied, we recommend to Ministers, who have always accepted our advice and who then issue Government approval." Derek Burke, Chairman of the Advisory Committee on Novel Foods and Processes, explaining how genetically modified foods gain approval in Britain. "We had no safety concerns [about genetically modified soya] and the Food Advisory Committee did not require labelling." Ibid *Government and the scientists it employs know best – but there is a risk communication problem.* "We used to think that all we had to do was to decide whether a novel food or process was safe or not, and a grateful public would accept what we said. We should have known better! Food irradiation, a process I and many others, believe to be safe is unusable because of fears connected with the word 'irradiation', which go back to the atomic bomb and are fed by concerns about nuclear power stations." Ibid
☺ **Individualist** "The new technologies are environmentally friendly and will lead to health benefits, an end to world hunger and reduced use of pesticides. 'There's no crop or person that cannot benefit. There's a tide of history turning. You can look back, or ask how you're going to feed the world,' Monsanto said." "Biotechnology is, and has always been, used to make bread, bacon, beer, wine, cheese, yoghurt, pickles and sauces. Humans have been manipulating plant and animal genes for about 8000 years, by breeding and cross-breeding. The difference is that, since Crick and Watson worked out the structure of the genetic code in 1953, it is now possible to work out exactly what is going on when an animal or plant grows faster, taller, or straighter, or withstands rust or blight or brucellosis." *if you can't prove its dangerous assume it's safe:* "Do you cease to approve all new technologies until everything you could conceivably imagine as a risk has been evaluated to the nth degree? ... I am confident it is safe. It is not possible to prove that it is entirely safe." Monsanto	☹ **Egalitarian** *abhors "unnatural" practices; is averse to unpredictability; fears technology dependence, and the polarising socio-economic consequences of the concentration of the ownership of the new technology in a small number of hands* "Robert Shapiro [CEO of Monsanto] ... has to find a market for the products his company has spent billions developing ... The wants and needs of ordinary humans are incidental. This 'growth at any costs' attitude on the part of the world's corporate giants is destroying not just our physical environment but the social environment that nurtures human community. ... The biotech industry [seeks] to prohibit labelling of genetically modified foods. ... The premium now is clearly on ignorance. ... Whatever the multi-million dollar spin merchants care to tell us, the scientists cannot guarantee their results. ... man's tampering with nature in this way is a recipe for disaster straight out of a horror movie. And you know what comes next. Nature fights back." *if you can't prove its safe assume it's dangerous:* "We cannot just release these things into the environment and hope for the best" Greenpeace

Figure 2. Ecological risks and prospects of transgenic plants: a typology of bias

Where risks are directly perceptible

everyone takes risks; *everyone* is a risk manager;

- taking risks leads, by definition, to accidents; the pursuit of a world free of accidents is a futile exercise;
- it is important to distinguish self-risk (e.g. eating too many cream buns, or beef on the bone) from behaviour that puts others at risk (e.g. unhygienic practices on a food production line); the second is a legitimate area for regulation, the first is not;
- attempts to criminalise self-risk are likely to be worse than useless; they are likely to redistribute the burden of risk in ways that harm innocent third parties;
- all genetically modified products should be so labelled to permit individuals to decide for themselves whether they wish to use them;
- risk management is a balancing act; institutional risk managers who do not take account of the reasons that people have for taking risks – the rewards of risk – will be frustrated.

Where risks are perceived with the help of science

- science can reduce uncertainty by illuminating the connection between behaviour and consequence;
- science, effectively communicated, can defeat superstition and purely imaginary scares, but
- science cannot provide "objective" measures of risk;
- risks come in many incommensurable forms that defy reduction to a common denominator;
- the act of measurement alters that which is being measured;
- risk is a reflexive phenomenon; in managing risks we are continually modifying them; in the realm of risk Heisenberg probably rules.

Where scientists do not know or cannot agree

we are in the realm of *virtual risk* where plural rationalities contend;
virtual risks are cultural constructs;
they may or may not be real – science cannot settle the issue – but they have real consequences;
the precautionary principle is of no help, different rationalities adhere to very different versions of the principle;
virtual risks are a fact of life; science will never have all the answers;
humility in the face of ignorance is a precondition for civilised debate about virtual risks.

Methods for Risk Assessment of Transgenic Plants
III. Ecological risks and propspects of transgenic plants
K. Ammann, Y. Jacot, V. Simonsen and G. Kjellsson (eds)

Dilemmas of risk-assessment research for transgenic crops

Les Levidow and Susan Carr

Centre for Technology Strategy, Open University, Milton Keynes MK7 6AA, UK

The political-scientific debate around biotechnology exemplifies the wider phenomenon of the "risk society". New technoscientific developments are widely perceived to impose unpredictable risks. At issue is the credibility and adequacy of present scientific knowledge for risk assessment. As experts overtly disagree, safety claims are opened up to greater public scrutiny. There ensue conflicts of accountability about how risks should be defined, calculated and managed [1].

For transgenic crops, precautionary research has sought to generate new knowledge which could resolve the risk controversy. Nevertheless it has been difficult to gain scientific consensus and public confidence for safety claims, especially for large-scale commercial use. The regulatory procedure responds to an evolving debate over the criteria for evidence of safety.

Amidst pleas that regulation be "based on science", there are contending accounts of how this should be done. These are crystallized in disagreements (or doubts) about the data which regulators "need to know", especially before granting market approval for a transgenic crop. Even after such approval, the commercial stage comes under pressure to be managed as yet another experiment. At issue is not simply "acceptable risk", but the acceptability and plausibility of potential *effects*.

There have been awkward attempts to fit the precautionary features into the "rational" stereotype of science-based risk regulation. This stereotype moves sequentially from hazard identification, to risk assessment, to risk management. At the "assessment" stage, the magnitude of hazard is multiplied by its likelihood or frequency; that result in turn informs the management decision about what measures are needed to minimize the realization of any hazard [2–4: p. 22].

Such a logical sequence presumes that hazards are objectively knowable. For transgenic crops, however, "likelihood" really means plausibility or predictability; thus "risk management" accommodates a predictive uncertainty about potential effects. At the same time, the term "magnitude" presumes that particular effects are unacceptable; such definitions of environmental harm influence the "hazard identification", thus reversing the stereotypical sequence. All these judgements are subject to change, even for the same product in the same country.

Not surprisingly, European safety regulation has had difficulty in overcoming the risk controversy. The EC Deliberate Release Directive 90/220 was intended to "establish harmonized procedures and criteria", especially for Europe-wide market approval of transgenic products. In judging such approval, however, EU member states have given different interpretations to key statutory terms, e.g. "adverse effects" and the "step-by-step" principle [5–7].

Un/acceptability baseline

As regards "adverse effects", each marketing application has led to disputes over what potential effects must be prevented. For identifying such effects, familiar reference points include: lower biodiversity, harm to non-target organisms, pest resistance, crop-protection methods, and overall herbicide usage. Within each reference point, disagreements arise over the acceptability or statutory relevance of specific effects. Some protagonists cite the effects of present agricultural practices as an acceptable baseline for evaluating environmental effects of cultivating transgenic crops. Other protagonists request evidence that a new crop will provide an environmental improvement and will not preclude any potential options for sustainable agriculture.

For example, transgenic oilseed rape may spread its herbicide-tolerance genes, *via* volunteers or hybrids which become weeds; their presence could preclude and/or encourage specific changes in agrochemical usage. Critics demand that any such scenario be evaluated as an "adverse effect", while proponents tend to regard it merely as an "agricultural problem" which lies outside risk regulation. So far the latter view has prevailed: market approval has been granted on the basis that farmers would still have other ways to control weeds. Research continues to test the hybridization capacity of *Brassicas*, yet such new knowledge serves to intensify (rather than resolve) the original controversy.

Another example is crops modified to produce a toxin from naturally occurring *Bt* microorganisms. Such crops may intensify selection pressure for *Bt*-resistant insects, thus eliminating a safe alternative to chemical pesticides. Critics regard *Bt* toxins as an indispensible tool for sustainable agriculture, while proponents foresee acceptable substitutes from other *Bt* genes or agrochemicals. So far the latter view has prevailed: market approval has been granted on the basis that farmers would still have other ways to control insect pests. Research continues to investigate and manage the causes of *Bt* resistance, yet new knowledge serves to intensify the controversy (as in the previous example).

As regards the "step-by-step" principle, the EC Directive was intended to help gather safety data through a progressive scale-up, to assess commercial use as the final "step", and thus to ensure long-term safety. Yet market approval has been beset with disagreements over how to define and predict "adverse effects". In designing further research, there are methodological difficulties in simulating large-scale commercial use [8].

As some critics label the commercial stage "an uncontrolled experiment", the authorities attempt to control and monitor its effects. Post-market monitoring has provided a useful compromise, though the scientific methodology then becomes subject to further debate. And it may be difficult to justify mandatory monitoring (or adequate resources) to detect effects which are not officially deemed "adverse" or which are regarded as implausible.

Predictability baseline

Beyond any statutory framework, the "familiarity" concept was devised to provide a common language for risk assessment, yet it has diverse meanings [5: pp 146–147]. Officially, it means to assess whether present knowledge and experience is adequate for regarding a transgenic crop as similar to a "familiar" organism whose behaviour is predictable. That consensual definition begs some questions: What *un*familiar features of transgenic crops warrant the effort to obtain additional knowledge? What more experience can overcome their unfamiliarity, or "gain familiarity", and thus enhance their environmental predictability?

Such questions underlie disagreements over criteria for adequate evidence. Some scientists cite known characteristics of plants – e.g. as "familiar to the plant breeder" – yet these provide no consensual baseline. Partly at issue is whether phenotypic traits of conventional plants may be regarded as adequately familiar: Given that some plants have naturally occurring resistance to antibiotics, herbicides, viruses, etc., what is their genetic basis? and their ecological role? What is their similarity to the traits being inserted into transgenic crops? How should such comparisons inform risk research and regulatory judgements?

For example, transgenic crops could spread their virus-resistance genes and confer a selective advantage to related plants, in turn disrupting weed-control measures and/or undermining natural biodiversity. To clarify such scenarios, research has investigated the presence of virus-resistance genes in wild populations, and how these may differ in transgenic crops [9]. Other research investigates the presence of plant viruses, to clarify whether they may exert relevant selection pressure.

Yet the empirical results are cited to propose further research before market approval is granted – or, alternatively, to declare that a virus-resistant plant is adequately familiar. More generally, when regulators claim to have found "no evidence" of specific effects, they come under pressure to look in more imaginative ways, or to present evidence that such effects could not occur. Thus the "familiarity" concept becomes a focus of further argument over the burden of evidence.

"Need to know"?

Concerning "Ecological Risks and Prospects of Transgenic Plants", the subtitle of this conference asks, "Where do we go from here?" As suggested above, any answer must rest upon specific models of nature and agriculture – e.g. of biodiversity, selective advantage, crop-protection methods, etc. Thus we also need to ask: What models of reality should guide risk research?

In sum, a value-laden science both informs and follows regulatory policy. Risk-assessment research links judgements about the acceptability and plausibility of potential effects. Scientific researchers face related dilemmas – e.g. how to link the normative and predictive judgements, how to simulate large-scale use, how to optimize the conditions for unexpected effects to occur, and how to justify resource-allocation for such efforts. The available resources will depend upon judgements about the data which regulators "need to know" – e.g. before relaxing initial controls, before granting market approval, or for confirming safety assumptions afterwards.

Indeed, we may say that risk regulation involves an implicit politics in defining the scientific uncertainties, why they matter, and the criteria for resolving them. The prospects for research funding will be enhanced if scientists publicly argue that market approvals should wait until greater scientific knowledge is acquired, though this stance would mean taking political responsibility for regulatory delays. Then, again, it would no less political to accept that present evidence is adequate (or even to remain silent). Such dilemmas arise from the pervasive conflicts over how to regulate transgenic crops.

References

1. Beck U (1996) Risk society and the provident state. *In*: S Lash et al.. (eds): *Risk, environment and modernity*. Sage, London, pp 27–43
2. DoE/ACRE (1993) *Guidance note no.1: The regulation and control of the deliberate release of genetically modified organisms*. Department of the Environment, London
3. DGXI (1996) A framework approach to environmental risk assessment for the release of genetically modified organisms, Doc XI/087/96, Revision 2
4. Kjellson G (1997) Principles and procedures for ecological risk assessment of transgenic plants. *In*: G Kjellson et al.. (eds): *Methods for Risk Assessment of Transgenic Plants II*. Birkhäuser, Basel
5. Levidow L, Carr S, von Schomberg R, Wield D (1996) Regulating agricultural biotechnology in Europe: harmonization difficulties, opportunities, dilemmas. *Science and Public Policy* 23(3): 135–57
6. Levidow L, Carr S, Wield D (1997a) Environmental risk disharmonies of European biotechnology regulation. *AgBiotech News and Information* 9(8): 179N–183N
7. Levidow L, Carr S, Wield D (1997b) Fausses notes dans le concert réglementaire. *Biofutur* 172: 26–28
8. Sweet J et al. (1997) The impact of releases of genetically modified herbicide-tolerant oilseed rape. *British Crop Protection Conference – Weeds*, pp 291–301. British Crop Protection Council, Farnham, Surrey
9. Cooper I, Raybould AF (1997) Transgenes for stress tolerance: consequences for weed evolution. *British Crop Protection Conference – Weeds*, pp 265–72. British Crop Protection Council, Farnham, Surrey

Discussion session 8: Conclusion, strategies, where do we go from here?

Werner Müller
As Les Levidow and others pointed out, a knowledge based decision has to value judgment. We first consider which effects are acceptable and secondly, the knowledge we have to reach our decisions to approve. Is the uncertainty too important or low enough? Has the opinion of organic farmers been considered? They rejected artificial fertilizer already in the 1920s, although there was not any risk assessment at that time. The organic farmers were not understood. It was only in the 1960s and the 1970s when environmentalists understood that overdoing artificial fertilizers will have negative environmental effects. The same happened with pesticides: Organic farmers rejected pesticides already in the 1930s and the 1940s, again there was no real risk assessment basis for this judgement.

Klaus Ammann
Basically I have nothing to object to here, Werner, except the things I have said before: Genetic engineering will reveal in future its great ecological potential, and wisely used there will be a time when organic farmers will have a difficult time explaining why they still exclude GMO's. And: Today we have a lot of risk assessment going on for GMO's, that is the important difference compared to the times of pesticide introduction.

Jim White
Unfortunately in the US we already had Bt resistance insects. And those insects have arisen by organic farmers or other farmers misusing conventional Bt's. Everyone, including people that use Bt in the garden, need to use chemicals in a respectful and safe manner, because they are very important, no matter who uses them. The first Bt resistance insects among others that emerged in fields are cabbage loopers in Hawaii, the fields there are surrounded by mountains, they used Bt every 3 days repeatedly. Also in New York, Bt has been misused or overused in greenhouses.

Angelka Hilbeck
If you call it a misuse if organic farmers uses a Bt insecticide three or four times, then what is it putting it permanently in transgenic crops?

Jim White
Any misuse of any pesticide could develop resistance without proper management.

Angelika Hilbeck
Yes, sure, but I mean if you call this a misuse, then it is a misuse to put it permanently in the plant.

Jim White
The US has introduced pesticide management strategies in order to avoid premature emergence of resistance.

Werner Müller
We have implemented our pest management program through our methodology. To kill pests is not our fist step. Our first step is to introduce a better rotation including seven or eight crops, not just only three of them. In many cases this will minimize harmful effects of insects or other pests so that if we need Bt we only need it in certain and very few cases in Austria. In Europe, we did not find any resistance problems with Bt up to now, because it is really rarely used, especially in potato. There are many farmers which do not use Bt against potato beetle.

Klaus Ammann
Well, I agree with everything except for you have not given me any shutting out argument for genetic engineering, not a single one. Still I believe that future GMO's will reveal a surprising ecological potential which should size the interest of organic farming. But again: I can accept all your valuable remarks and I think those emphasizing genetic engineering as the sole tool for the future in order to achieve a more intensive agriculture feeding the world, they should learn from organic farming strategies, and that is what I mean by working together.

Phil Dale
Les Levidow, one of your messages was that science of risk assessment is value laden. But all science decision making pesticide, what ever, is influenced by our personal values or society values. So is it surprising that our risk assessment is also value laden?

Susanne Lauber
I would like to come back to the point about ideology, whether the organic farmer should use genetech, yes or no. And I really like the discussion we have had in the evenings about who is ideological and who not. And finally, I think we have all our ideologies and the point is important about where we do meet, where we can stay maybe with our own ideologies, but where have we found common ground. And I think we cannot find this common ground now. Returning to E. Schulte, she showed us in Switzerland that we have 7% organic farmers, 38% conventional farmers and 55% integrated production. And if you take the integrated production, I think this can be how we can feed the world. Organic farming is not the solution for all over the world, maybe it is interesting for Switzerland, but when farmers learn from each other how to plant, how to cultivate and maybe also how to get better seeds from genetic engineering, this would really be the common ground and each one could stay with his own view.

Klaus Ammann
I still have to say one thing about organic farming and genetic engineering coming together. I can easily understand that from the point of view of marketing it is not the time. I mean, organic farmers need to develop their market and they have a hard time, and if they are just overthrown and invaded in a way which can truly threaten the market, then I think it is not adding to the peace we need.

Jens Soth
I am awfully sorry to respond to this discussion with a little argument again, because I see the discussion should go in another direction, so later on I will contribute to our general discussion. I have exactly a contradicting hypothesis. Organic farming feeds the world. We have conducted a study and internationally we found only twelve crops where we could not find examples where organic farming outperforms in its yield the normal conventional way of farming. That is not scientific, I know, because it is only singular examples. But they are there.

I found John Adams' idea very brilliant, but risks are often socialized and benefits, such as the ones from selling transgenic seeds are privatized.

Andreas Seiter
I agree, but consider the following: Fact is of course the chairman of Novartis makes significantly more money than most people here in the room, but the vast amount of the money the companies earn is going back to the shareholders and the shareholders are, for example, pension funds of universities and the shareholder is yourself if you go to a bank and ask for 5% interest rate instead of 3%. And the other point is that people are making money from jobs which are provided by industry which is profitable.

Going back to organic farming, I see a similar dilemma here. It is clear that certain risks from pesticides and potential risks from transgenic crops are not incorporated in organic farming concepts. They are avoided. But what about the productivity of the area, the land used here, which is for me a major ecological impact and what about the per ton harvest impact of fertilizer, if you use natural fertilizers compared to integrated farming practices. I'm not so sure that in an overall assessment on a worldwide basis that the equation is still the same.

Richard Braun
We just should not quite forget the farmers in all this discussion in the US; it is very clear that the use of Roundup Ready Soja increased by a factor of five or six from 1996 to 1997. So there were something like five times more farmers willing to buy this material and of course they did that for their profit. They clearly think economically.

Jens Soth
Of course, Mr. Seiter, you are putting difficult problems on the table about shareholder values and what we define at our institute as the sustainability of capital markets. But I can assure you that this issue which will be taken up in future by a lot of people and it is a very important from companies' points of view to think about it. Although I do very much agree with you that it is a problem for you and it cannot easily be solved, but there are a few approaches to how to solve it. Actually in Switzerland we have rather good approaches for ecoperformance portfolios even on capital markets.

Jens Soth

And to argue again on organic farming. Yes, you are right. One has to look at the productivity per landscape and I come from a institution where we do life cycle assessments; we ought to put into question what is the input into any system? And again, we see a lot of advantages from organic farming. Just if you calculate, for instance, energy efficiency. What you put into the conventional system and what you get out is in relation to organic farming not very competitive. But that needs of course a very broad prospective and holistic approach.

Klaus Ammann

Well, the wrap up now still contains a few decisions and conclusions and resolutions, but my feeling from the statements of the keynote speakers is that we have plenty of material from which to distill out something, and because I did not get any negative feedback about the statements of the keynote speakers, I would now ask, are there any additions from anyone about what may have been missed by the keynote speakers and the rapporteur statements?

Les Levidow

Many speakers have proposed methods for monitoring, but these proposals were not connected to specific harmful effects with an argument that society must evoke the resources to monitor, because such and such effects are unacceptable. If such an argument is not made then the scientific technical capability either will not be developed or the resources were not to be found. And more importantly, products will be commercialized which may have such effects by which time it may be too late. So far there has been only a vague link between these scientific proposals on the one hand and the regulatory judgments about potential effects.

Simon Barber

I want to make a comment about concepts of effects. I think we can do a very good job of monitoring for potentially anticipated effects. We can use science to anticipate effects. And as such, with certain modified plants, we can anticipate the sorts of effects that might result and develop in a monitoring scheme forward. We cannot possibly design anything meaningful for unexpected effects. We cannot look for unexpected effects. We do not know what we are looking for. I mean, we are all talking about Bt resistance insects. So we should monitor for it. A question for Les: I find his arguments very interesting, but I never find a solution to the issues he has raised. I have not seen alternative models. It is very difficult to know how to react. It is very easy to pick holes in a thing as it exists. But it is much easier to improve things if you are given good concrete suggestions for improving models.

Christian Damgaard

It is to monitoring contra risk assessment. The one thing that struck me by, I think, a very nice presentation by Peter Kareiva, was that the x-axis was 20–40 years, so the monitoring has to pick up a signal, the monitoring has to go on for at least 10 or 20 years and many of the commercially interesting crops are only grown for ten or 15 years. So any result of the monitoring would according to these perspectives come so late that the crop monitored is already replaced by another one. I think that it would be too weak for that reason alone to say that monitoring has a very important effect. I think that risk assessment before the crop is released has first priority. And the risk assessment being a probability assessment from the scientific community multiplied by the effect which has to be defined by the regulators as a summing up of the political and social discussion.

Peter Kareiva

Maximum sample here was less than 1 of 100 percent of the landscape. I was going for a very small coverage and if you emphasize the coverage, if you want an answer within 2 or 3 years then that means you are going to find that sampling tends to 20% of a landscape. So you can answer that question with the monitoring program.

Christian Damgaard

Those very nice examples by Jarle Tuftos in the modeling session, he showed that after a 5-year period there was absolutely no difference between a positive selected gene and a negative selected gene of trees. So if you want to monitor transgenic trees you have to allow for some time for the effects to occur. After 2, 3 or 5 years there were no significant effects and differences between positive and negative selected genes.

Phil Dale

I just want to note something that Simon said about monitoring. Clearly, we can have more planning in monitoring for things that we can anticipate. But Simon's comment was it is impossible or difficult to monitor things that we cannot anticipate. Again, I think that if we involve farmers who have thousands of hectares, they are technologists and scientists in their own right, and I think we should bring them much more into the debate and involve them in it. They are on the front line in such a way as to notice these things.

Jens Soth
I am happy to present a pragmatic solution for monitoring, but there is one hook. It does not help the regulators. So the solution is, or one approach for a solution is, that the companies pick up monitoring in a way that they have got a certain responsibility. And according to proactive definition of companies, they should take over responsibility. I am very much in favor with Philip Dales' ideas about integration of farmers, so the practicable idea would be really also to integrate the response that Tom Nickson is getting from the farmers also to make this response open. You can shape a monitoring of the more qualitative and soft parameters like soil erosion, socio-economics effects, but I see it is not a help for the regulators.

Klaus Ammann
Well, actually it could well be a help also for the regulators. I cannot see any contradiction and it is no paradox for me that these farmers' relationships with the American Biotech companies are much more intense than in Europe, because these are farmers making their own decisions. Monsanto takes good care of these farmers and the seed companies. It is in their own interest. I would like to emphasize what Phil Dale said, it is important to involve the farmers, involve the biotech companies, but here we are in a dialogue between science and the biotech companies and there I would like to make a possibly audacious remark: The risk assessment research community could also be involved by the biotech companies in the future when it comes to new products and visions, because actually to put in slightly rude words, the risk assessment community always has to run after the new products. I know that here we are digging into the sensitive competition area for the companies. But still my dream would be at least to enhance this dialogue of the present 2 days for the near future.

Björn Åge Tømmeras
I think I fully agree to highlight the questions about monitoring. I think maybe it could be a good idea to make a little step forward. It is especially important for a dialogue between the conflicting partners including NGO's or other organizations. It should include the concern of biodiversity.

Klaus Ammann
Well, if I may just answer here, because it is biodiversity directed. There are bridges to be built where biotech companies, regulators and risk assessment researchers could work together, that is the method of biogeographical assay in Europe. Think about all the old crop cultivars where we do not have a clear biogeographical picture. We have only a very scanty idea of the genomic resources of old cultivars in Europe and that could easily be linked with the mapping of red list, pink list and blue list species. Blue list species are those which are definitely on the safe side through conservation projects. Pink lists comprise species on the edge of becoming red list species.

Jim White
I would like to support Phil Dale's suggestion that farmers be involved. In California, which is our largest agriculture producing state, the majority of farmers have bachelors degrees. They are highly educated people. I think they need to be involved. In the US, also, we have over 100 000 USDA federal employees in agriculture and they are located in every locality across the US. They should be involved, every state has its own agricultural system located in every locality. And all those people are good sources to be involved in this monitoring and they already are involved in monitoring for a variety of pest reductions such as looking for mad fly coming into California. I would like to make one clarification about what I said previously about this free number from Monsanto. This was an additional thing for the transgenics. Like in Europe, you can always call your state or agriculture official to complain. And they do, but this free number was something new to respond to the use of the first transgenic plants in potatoes.

Martin Keller
I totally agree with the statement you made that we need a risk assessment specific for transgenic plants before deregulating them and this will be a help for regulatory agencies. On the other hand, we may need monitoring to answer the big questions that you asked about biodiversity geneflow. But I do not see any answers for those questions in a monitoring specific to transgenic crops. In those talks I heard nothing about long-term problems restricted to transgenic plants and not to any other way of shipping cultivars and seeds around the world. So I would suggest to concentrate on a risk assessment on transgenic crops which is based on the knowledge we have today for

possible risks and this will be the basis for deregulating the crops. A monitoring should not be focused on transgenic crops but on agricultural practice in general.

Les Lewidow

Is it often said that farmers are capable of noticing things if something goes wrong and, of course, they will notice anything which harms their crop, they may even notice effects other than harm to their own crop. But many potential effects may not be noticed by farmers. So this is not a perfect solution. For example, if insect pests develop resistance to Bt, eventually this will be noticed, but the resistance may be developing for a long time before the crop damage is noticeable and by then the problem may be serious. If Bt crops harm non target insects this may never be noticed by farmers. And we could go on and on with the list, so: Yes, it is fine for risk assessment research to include farmers, but it is also necessary to think about types of monitoring before commercial use and during commercial use which go far beyond anything that farmers themselves can notice.

Phil Dale

I agree that it is not the total answer. Science must be injected into monitoring. But I think what you just said is underestimating the ability of many farmers. They know about more than just the pest on their crops and the diseases. They are aware of skylarks, they are aware of many aspects of the environment. So, just to think that farmers are there for profit, there for getting as much out of the land as possible, I think is in a way insulting to them. They are aware of the problems and many have been born on the farm, and their grandfathers, too.

Klaus Ammann

I fully agree. Farmers by their own experience are holistically thinking ecologists. And those who do not think in these categories will not be successful farmers.

Jim White

I would like to state this Hawaii story where the first Bt resistant insects were found in the field. The story, as I have been told by the Hawaiians, is that the farmers noticed the Bt was becoming less and less effective, they had to add more and more Bt. They called their state person who encouraged an entomologist at the Universisty to come and investigate it. So I think in this case, as Phil said, it has worked in the first situation and they are looking at their ecosystem.

Klaus Ammann

Yes, I have worked for 5 years in a big monitor program of Switzerland mapping the Swiss flora, and this is scientifically not very rewarding. I ended up in one little mention in the introduction of two thick volumes for 5 years of scientific work, but what I learned in this period was how to approach problems with a biogeographical assay. It is an early warning system and it is the best we have, although we calculated something like a 75% precision in getting distribution data.

I think it is time to finish the discussion in session 8. As a summary, we really have focused on this monitoring and I would like to wrap up this conference by mentioning a few things which come to mind now.

First, we have had as an introduction the fantastic Encyclopedia Danica on all aspects of "nice to knows" and "need to knows".

Then two modelers from the US and UK gave us lots of very valuable hints of how to structure monitoring and anything else would be a misunderstanding.

There were many other contributors, I should not mention names, but I should mention the sound science coming from the Island over the channel, and for me, I had lots of fun.

Also I highly appreciated the collaborative atmosphere in the dialogue between corporate life science companies, regulators and scientists.

And also, I had lots of fun with some of the more critical comments coming from the Nordic countries. This is something like a cultural difference between the north and the south which has been clearly demonstrated here in the conference.

Overall, I think we have made some progress and we tried our best to sort out the main highlights with summing up the discussions, the rapporteurs have given a clear picture on priorities which should be met in the future.

When it comes to the conference proceedings, it would be nice and easy to fill three volumes with all the tapes comprising 8 hours of discussion, but we will try to distill out whatever we can.

Next Tuesday, we have a press conference where we have to concentrate on a few pages delivered to the press. Thank you all. I have learned a lot and I hope that this is something you can say also from your own perspective.

Vibeke Simonsen, Yolande Jacot, Pia Rufener Al Mazyad and Klaus Ammann

Press release
An international high level Symposium including scientists, regulators and delegates from life science companies was organized in Bern in order to initiate an open discussion on the safety of field releases of transgenic crops. Participants from Europe, North America and Asia were chosen according to their professional activities in the field of risk assessment. In three days a collaborative learning process has been structured in such a way that, in a stepwise and argumentative proceeding, the most important research fields have been covered and in a final block some decisions about lacunes in risk assessment and future activities have been taken.

- In a first block we concentrated on the ecological impact of single transgenes as a basis for the next block discussions. All transgenes which can cross out should be evaluated according to their possible ecological impact.
- In a second block another important scientific basis for future risk assessment research has been presented: Modelling could be part of the solution: Complexity of ecosystems may become more transparent with the refinement of modelling methods.
- In block 3 many aspects of different time scales have been discussed: Long-term and short-term effects need to be addressed with different methodological approaches.
- In a fourth block basic questions about monitoring have been asked, new methods have been presented.
- A fifth block has been devoted to the complex questions of population genetics. Transgenes offer new opportunities for getting a clearer picture in risk assessment.
- In a sixth block decision-making processes have been discussed in the framework of regulating transgenic crops. Special emphasis has been put on the questions of harmonizing the regulation process.
- In a seventh block lacunes of risk assessment research have been discussed. Long-term monitoring has been determined as one of the most important lacunes.
- In a last block decisions have been made in order to answer the most important question asked by the conferees: Where do we go from here? Without any opposition from the conference participants it was decided with a clear priority that the establishment of a long-term monitoring would be very important for the future.

Methods for Risk Assessment of Transgenic Plants
III. Ecological risks and propspects of transgenic plants
K. Ammann, Y. Jacot, V. Simonsen and G. Kjellsson (eds)

Bern conference on gene flow: one scientist's reflections

Thomas E. Nickson

Monsanto Company, 700 Chesterfield Parkway, North St. Louis, MO 63198, USA

The debate that surrounds the release of genetically modified plants into the environment is rich with information, misinformation, and opinion. This conference was a worthwhile attempt to bring these factors into a dialogue with basis in objective science. The goal was to summarize the current status of the science and propose tangible actions for the future. Key scientists with a great deal of experience in evaluating transgenic plants were given the opportunity to summarize their work and engage in informal discussions. Complementing these presentations were presentations given by regulatory officials who have the responsibility to assess the data and make decisions based on the risks present. I personally experienced many discussions, and obtained some valuable insights. It remains to be shown whether progress was made toward establishing the balance between risk assessment and pure scientific inquiry.

Possibly because genetic engineering is a new technology embroiled in public controversy, much of the ecological scientific community represented in Bern is confronting questions that are difficult to balance. Fundamental to this balance is understanding risk as a personal choice *versus* risk in a regulatory context. When asked how to assess the risk of a plant (not necessarily a genetically modified organism (e.g. GMO) in the environment, qualified scientists will design a series of experiments which take years to complete. These experiments will be detailed and based on their personal interpretation of the questions to be answered. Furthermore, the end point of pure science is the next question, rarely is a definitive answer obtained. This process creates a dilemma for the regulator and an economic problem to the industry that expects return on their investment. The discussion at the Bern conference highlighted the fact that for regulatory purposes there is "need to know" data, that which is critical to making a regulatory decision. Much of the current debate on the risk associated with transgenic crops is centered on detailed characterization of the science.

John Adams probably summed the issue best in his discussion of the nature of risk. Risk, as it is understood in science, cannot be quantified. As such, the scientific method, while useful to answer specific questions, is not able to break the personal biases that shroud acceptable risk. Despite its scientific basis, the conference was confronted with the basic problem of breaking personal paradigms with objective fact. In addition, the knowledge of ecology as it applies to agriculture may be perceived by some as insufficient at this point of time. It is in this area, the coalescence of agricultural and ecological sciences, that offers opportunities to better understand significant risk in agriculture, and allows the implementation of new food production technologies for the future. The conference highlighted this need.

For the short-term, regulators in Europe will have to work with key scientists committed to objective results and developing the data needed to make regulatory decisions. Appropriate methods to monitor the commercial release of modified plants seems to be a consensus outcome of this conference. Appropriate monitoring would involve farmers as stewards of their land, industry as stewards of their products, regulators and scientific experts in agriculture and ecology. No group should be excluded as all are major stakeholders.

Methods for Risk Assessment of Transgenic Plants
III. Ecological risks and propspects of transgenic plants
K. Ammann, Y. Jacot, V. Simonsen and G. Kjellsson (eds)

The concept of familiarity and its role in the commercialization of pest resistant genetically engineered plants

James L. White

Senior Operations Officer, Biotechnology and Biological Analysis, Animal and Plant Health Inspection Services, U.S. Department of Agriculture, Unit 147, 4700 River Road, Riverdale, MD, 20737-1237, USA

In 1987, the National Academy of Sciences [1] in the United States published a report entitled, "Introduction of Recombinant DNA-engineered Organisms into the Environment: Key Issues," that stated that safety assessment of a recombinant DNA modified organisms "should be based on the nature of the organism and the environment into which it will be introduced, not on the method by which it was produced." Following this report, the U.S. Department of Agriculture (USDA), the Environmental Protection Agency (EPA), and other regulatory agencies asked the National Academy of Sciences-National Research Council (NRC) to evaluate scientific information relevant to making decisions about the field testing of engineered plants and microorganisms.

In 1989, NRC [2] published its report entitled, "Field Testing Genetically Modified Organisms." One conclusion reached was that the use of plants modified by classical breeding techniques for field testing has a history of safe use, and that crops modified by engineering should pose risks no different from those modified by classical genetic methods for similar traits. Thus, if the genetically modified plant is phenotypically similar to a plant that has been (or could be) bred by traditional breeding techniques this parallel association is called familiarity. The concept of familiarity allows regulators to draw on past experience with introduction of modified plants into the environment. Familiar does not necessarily mean safe. It does mean that the level of risks associated with the introduction of new pest resistance genes into plants by classical methods and the evaluation of new cultivars by national variety registration agencies, has made the introduction into the environment of these types of modified plants of negligible risk. Other important familiarity factors are whether the plant is new to the particular environment where it is intended to grow, the nature of the trait (gene), and that the evaluations should be made on a case-by-case basis. All engineered crop plants that have been commercialised in the U.S. to date have been grown in the same environment that their non-modified progenitors were grown in. (For a more detailed discussion of the concept of familiarity see an article by Dr. Simon Barber in this volume [3]).

So how did the USDA use the concept of familiarity in its reviews of the commercialisation of transgenic plants? One of the more controversial commercialisation reviews in the U.S. was for the approval of virus-resistant squash (*Cucurbita pepo*) plant. This was the first approved transgenic virus resistant plant and first transgenic plant approved for use in the country where the ancestral progenitors of the crop are found. Asgrow Seed Company (now Seminis Vegetable Seeds, Inc.) had engineered the squash to contain the viral coat proteins of zucchini yellow mosaic virus and watermelon mosaic virus 2. These are two of the six or more major virus diseases that effect cucurbit production in the U.S. and many other parts of the world including the Mediterranean basin. Critics suggested that movement of virus resistance transgenes from commercial squash to wild squash would increase weediness of the wild plants. As key but not the sole component of the USDA's review it was noted that phenotypically-identical, virus-resistant squash plant developed by classical breeding techniques was already being sold by a competitor. Thus, the commercial sale of the classically-bred virus resistant squash would need no formal written assessment, just the traditional assessment that any new cultivar receives before sale performed by plant breeders and/or scientists involved with variety registration. USDA concluded in its risk assessment of the engineered squash, "... that ZW-20 squash will be just as safe to grow as virus-resistant squash cultivars developed through traditional breeding practices". With some virus resistance transgenes that are derived from viral sequences like the coat protein used to develop the Asgrow squash, some potential issues have been postulated. These issues include transcapsidation, synergy, movement of subliminally-infecting viruses, and recombination. Those issues were discussed in USDA assessment [4] and general discussion of these issues have been published by the OECD [5].

As pest resistant transgenic plants reach the commercialisation stage world-wide, using the knowledge gained from the use of new varieties of plants that contain pest resistance genes introduced by classical breeding may be quite valuable in facilitating the evaluations of transgenic plants that will be undertaken in each country.

References

1. National Academy of Sciences (1987) Introduction of recombinant DNA-engineered organisms into the environment: Key issues
2. National Research Council (1989) *Field testing genetically modified organisms*. National Academy Press
3. Organization for Economic Cooperation, Development (OECD) (1992) *Safety considerations for biotechnology*. (Available electronically at: http://www.oecd.org/ehs/public.htm)
4. United States Department of Agriculture (USDA), Animal, Plant Health Inspection Service (APHIS) (1994) Petition 92-204-01 for determination of nonregulated status for ZW-20 squash. environmental assessment and finding of no significant impact. (Available electronically at http://www.aphis.usda.gov/biotech/not_reg.html)
5. Organization for Economic Cooperation, Development (OECD) (1996) Consensus document on general information concerning the biosafety of crop plants made virus resistant through coat protein gene-mediated protection. Series on harmonization of regulatory oversight in biotechnology. No. 5. (Available electronically at: http://www.oecd.org/ehs/public.htm)
6. National Academy of Sciences (1987) Introduction of Recombinant DNA-Engineered Organisms into the Environment: Key Issues. National Academy Press, Washington DC, 24 pp
7. National Research Council (1989) Field Testing Genetically Modified Organisms: Framework for Decisions. National Academy Press, Washington DC, 170 pp

Poster session

Methods for Risk Assessment of Transgenic Plants
III. Ecological risks and propspects of transgenic plants
K. Ammann, Y. Jacot, V. Simonsen and G. Kjellsson (eds)

Assessment and management of field testing of transgenic crop plants in east asian countries

Akira Hasebe

National Institute of Agrobiological Resources, Kannondai,Tsukuba, Ibaraki 305, Japan

In China, virus resistant tobacco has been field tested since 1991, and it has been used in industrial tobacco manufacturing for national consumption since 1992. The cultivated area exceded 1 million ha corresponding to 5% of total tobacco plantation in China. At that time, China had no biosafety regulations. This means the field tests were carried out without national control. Recently, Chinese government issued safety guidelines for field testing of transgenic crop plants. Thereafter, all field tests of transgenic crop plants are being reviewed based on the guidelines.

In Japan, several guidelines for industrial use of rDNA organisms have been established since 1986. Among these guidelines, environmental release of transgenic plants is mainly regulated by MAFF (Ministry of Agriculture, Forestry and Fishery) guidelines. There are long steps before commercial release of transgenic crop plants in Japan. Three stages of safety evaluations exist before commercial release of transgenic crop plants. At the first stage, transgenic plants should be cautiously kept in laboratories or closed greenhouses, and safety evaluation such as residuality of *Agrobacteium* as vector, morphological characteristics, pollen form, fertility of pollen, longevity of pollen, pollen dispersal by wind, production of toxic substances from root exudates is needed before moving to the secons stage. At the second stage, transgenic plants can be maintained in ordinary greenhouses, and following safety evaluation tests are required before moving to the third stage: Reproductive characteristics such as fertility of seeds, seed germination; Toxic substances produced in stems and leaves; Influence of succeeding crop; Influence of soil microflora. At the third stage, transgenic can be planted in small scale fields described later, and ecological impacts are estimated before commercial application.

First field trial of transgenic crop plant in Japan was carried out in 1991. A research team of the impact of the release of the transgenic tomato on the environment was organized in 1989. More than 10 scientists including soil microbiologists, plant pathologists, entomologists, population geneticists, plant molecular biologists, plant physiologists, and plant ecologists joined the team. The following characteristics were compared between the parent and the transgenic tomato by the team. (1) Morphological and growth characteristics, (2) Reproductive characteristics, (3) Evaluation of harmful impact to other plants, (4) Influence on soil microbial flora, (5) Weediness, and (6) Residual *Agrobacterium* as vector. There was no difference between transgenic tomato and the parent tomato except the introduced traits. Actually, cultivation was safely carried out and negligible impact on the environment where the transgenic tomato had been cultivated was detected. The transgenic tomato was approved to be cultivated in an ordinary field in the following year.

As followed by the first field trial of transgenic tomato, more than 60 field trials of transgenic crop plants including rice, soybean, rapeseed, corn, potato have been conducted in Japan since 1991. Safety assessment of those released transgenic crop plants has been carried out in the same manner as done in the first transgenic tomato in 1991. All field tests of transgenic crop plants have been carried out in the designated isolated field sites in Japan. The designated test sites have several measures to assure safety for human and environment. First, the test sites are surrounded with fences to prevent access by other people who are not involved in the tests. Secondly, incinerators are installed to use when some unfavorable accidents might occur. Thirdly, all machinery used for agricultural practice is washed and disinfected on site.

Methods for Risk Assessment of Transgenic Plants
III. Ecological risks and propspects of transgenic plants
K. Ammann, Y. Jacot, V. Simonsen and G. Kjellsson (eds)

Mechanism of DNA integration into *Agrobacterium* genome-its relevance to horizontal gene transfer

Kornel Burg, Karin Hohl and Maria Berényi

Austrian Research Centers, A-2444 Seibersdorf, Austria

In the present work, we investigated the putative transfer of the foreign genetic information present in transgenic plants into soil micro-organisms.

We set up an experimental model system involving the construction of a transgene potato plant with a marker gene and an acceptor organism in which the appearance of this information is readily detectable. We choose *Agrobacterium tumefaciens* as an acceptor organism since it is a common soil micro-organism in nature and also capable to distribute DNA by transformation. As a marker the chloramphenicol acetyltransferase (CAT) gene with bacterial promoter was chosen. In case of transformation its presence can be easily detected by the appearance of resistant bacteria against chloramphenicol (Cm) in the *Agrobacterium* population. Furthermore we also provided homologous sequences to the transgene information in the *Agrobacterium* to test the role of putative homologous recombination events which may facilitate the integration of the foreign genetic information.

We applied two different transformation systems either initiating the DNA transfer by freezing and thawing the *Agrobacterium* or by electroporation. The former system provides relative low efficiency while electroporation proved to be an effective transformation system. Freezing and thawing of *Agrobacterium* also occurs in nature so we considered this transformation system as a model for natural conditions. Electroporation is made at low ion concentration what is rather uncertain in nature. Apart from these artificial transformation systems we also co-cultivated the transgenic plants with the *Agrobacterium* under *in vitro* conditions in order to find any natural mechanism which results in gene transfer from the transgene plant into the *Agrobacterium*. On the basis of our experiments we may conclude, that:

- *Agrobacterium* can be transformed under laboratory conditions with supercoiled plasmid DNA.
- *Agrobacterium* is capable to integrate DNA sequences similar to those found in its T-DNA by possibly homologous recombination.
- Linear DNA could also integrate into the *Agrobacterium* genome
- Single-strand DNA was also capable to integrate by possibly homologue recombination.
- No DNA transfer (either direct integration or by homologous recombination) could be detected from transgenic plants into *Agrobacterium* where the Agrobacteria were in contact with the plant roots.

Gene flow in selected swiss crops and related weeds, risk assessment for the field release of GMO's in Switzerland: case of wheat and oilseed rape

Roberto Guadagnuolo[1], Dessislava Savova Bianchi[1], Julia Keller Senften[2], Pia Rufener Al Mazyad[2], Yolande Jacot[2], Klaus Ammann[2] and François Felber[1]

[1]*Institute of Botany, University of Neuchâtel, Chantemerle 18,CH-2007 Neuchâtel, Switzerland*
[2]*Botanischer Garten, Universität Bern, Altenbergrain 21, CH-3013 Bern, Switzerland*

Possible gene flow between *Triticum æstivum* L. (wheat), *T. spelta* L. (spelt) and *Aegilops cylindrica* Host., as well as between *Brassica napus* L. ssp *napus* (oilseed rape) and *Raphanus raphanistrum* L. is studied in order to evaluate the risks of cultivation of transgenic crops in Switzerland. It is known that wheat and *Aegilops*, as well as oilseed rape and *R. raphanistrum* can hybridise, and the gene escape from cultivated species to the wild is possible.

This poster presents the preliminary experiments on establishing of diagnostic markers for each of the studied species. Distinctive morphological characters were defined. For *Triticum æstivum, T. spelta* and *A. cylindrica* in addition to morphology three types of genetic markers were determined, which allow to differentiate these species – isozymes, RAPD and microsatellites. For *B. napus* and *R. raphanistrum* only morphology and isozymes were investigated.

Two enzyme systems (Glutamate oxaloacetate transaminase and Peroxidase), 8 RAPD primers and 12 pairs of microsatellites primers distinguish clearly the three *Triticaceae* and three enzyme systems – Esterase, Superoxide dismutase and Malate dehydrogenase determined markers for both *Brassicaceae*. Genetical analysis were performed on plants of *A. cylindrica* and *R. raphanus* (growing in the border of oilseed rape field), collected in the wild in order to detect hybrids and evidence of introgression.

Experimental hybridisations for *T. æstivum, T. spelta* and *A. cylindrica* in the greenhouse and in the field were carried out. The progeny will be analysed, using these set up markers for the presence of hybrids and the rate of gene flow between the cultivated plants and their wild relatives will be determined.

TA-project "genetic engineering, breeding and biodiversity"

Rolf Meyer and Arnold Sauter

Office of Technology Assessment of the German Parliament (TAB), Rheinweg 121, D-53129 Bonn, Germany

The Office of Technology Assessment of the German Parliament (TAB) investigated in this TA project how conventional and genetically-engineered plant breeding can influence biodiversity and what is the resulting need for political action by the German Parliament. Investigations were restricted to the agricultural sector in Germany, taking into account European framework conditions.

An attempt is being made to systematically identify potential impact chains resulting from the introduction of new varieties produced using conventional or genetically-engineered plant breeding methods. At the three biodiversity levels – genetic diversity, species diversity and diversity of ecosystems – a differentiation is first made between direct and indirect impacts (Tab. 1). Only a limited amount of empirical knowledge is on hand to describe the impact chains.

Table 1. Potential impacts of new varieties on biodiversity

Level	direct impacts	indirect impacts
genetic diversity	diversity of varieties, genetic diversity between species	genetic diversity of the surrounding fauna and flora
species diversity	cultural species being cropped	soil organisms, auxiliary field flora, plant diseases, pests and beneficial insects
Diversity of ecosystem	crop rotation, agricultural land use	agro-ecosystems, related ecosystems

The general finding is that introducing transgenic plant varieties, compared to conventionally bred plant varieties, will not lead to any fundamentally new impacts on biodiversity in the medium-term. On the other hand, genetically-engineered plant breeding alone is not expected to make any significant contribution to conserving plant genetic resources, because the former impact chains will continue to exist.

A critical discussion is called for on the use in transgenic crops of biological antigens such as *Bacillus thuringiensis* toxin and virus sequences which do not normally infest plants. Particular attention should be paid to the development of resistance or formation of new pathogens, and the consequences this will have on farming practice and biodiversity.

Finally options for the conservation and sustainable use of biodiversity and plant genetic resources are developed and discussed.

Methods for Risk Assessment of Transgenic Plants
III. Ecological risks and propspects of transgenic plants
K. Ammann, Y. Jacot, V. Simonsen and G. Kjellsson (eds)

Final summary of the conference

Klaus Ammann, Yolande Jacot, Gösta Kjellsson and Vibeke Simonsen

Introduction

A conference sponsored by Swiss Governmental agencies and Industry, was held in Bern in order to enhance communication between science, biotech companies and regulators. More than a hundred participants from all over the world took part in this conference; among them were some of the most notable risk assessment researchers, as well as important representatives from biotech companies and some major regulatory office members. Views and knowledge were exchanged in an excellent collaborative learning atmosphere. The process was accompanied by lively discussions (some 10 hours were recorded), with ample time to clarify misunderstandings. The most important results produced during the conference are summarised below.

At a time when transgenic plants will soon be spread all over the world and commercial releases will be normal, we need to come to terms with monitoring the follow-up processes in the environment. Experts, regulators and biotech company delegates all agreed at this conference that field releases will take place in most regions of the world and that there is no reason to assume that transgenic plants *per se* are imposing a dramatic risk to the environment. But a combination of transgenes, the introduction of multiline genomes and effects stemming from large scale release of transgenic cultivars could well justify the establishment of a long-term monitoring.

Benefit of transgenes

The use of transgenes increasing herbicide tolerance or disease resistance may reduce the use of environmentally objectionable herbicides. The use of biodegradable pesticides may also reduce the number of pesticide sprayings, which saves energy and labour and considerably reduces the pressure on the environment. Genes changing quality traits may be useful for the improvement of food quality for humans and animals. Genes changing stress tolerance, e.g. tolerance to cold, salt or drought, may allow cultivation of plant species in new areas. The production of compounds alien to plants such as pharmaceutics or bioplastics can be an alternative to the production with the help of transgenic animals.

Escape of transgenes

It has to be considered that a transgene can be spread to other varieties of the same plant species or to related plant species by hybridisation and introgression, i.e. incorporation of the transgene into the genome of the related species. In order to avoid escape of alien transgenes encoding, for example, pharmaceutics or bioplastics, we need geographically isolated production systems, which may also require physical or biological containment.

During the conference, it was stated that not only transgenic plants may have negative effects on ecosystems, but also conventional plant breeding and cultivation procedures have had and will have an impact on the environment. Gene flow to related wild species is a long-known phenomenon and not a novelty of transgenic plants; basically it is essential for species evolution. We need to distinguish between risk and truly negative impact. There was not a single case of negative impact demonstrated in the symposium, but everybody agreed to remain vigilant.

The conference unanimously stated that invasive species are a big threat to specific environments and should be taken very seriously, although there is not necessarily any connection with genetically engineered plants.

When evaluating the risk for spread of transgenes, two components have to be taken into account: the plant species and the transgenes.

For the plant species, it is important to know how easily it can hybridise with other species; especially in centres of species origin, great care is essential. For the transgenes, we need to know the effect which they may have on the environment either in the transgenic plant or with contamination of related species. Both the plant species and the transgenes have to be assessed, and a risk assessment on several levels needs to be performed.

Risk can be expressed as follows: risk = probability × hazard,

where probability means the chance transgenes can outcross to wild relatives, or that the transgenic plant itself spreads into nature. Hazard is the negative effect of the transgene or the plant on the ecosystem. Both probability and hazard have to be evaluated *case-by-case*, as both of these parameters are influenced by the ecosystem in which the transgenic plant grows. Nevertheless, the principle of *familiarity* may be applied to several cases, e.g. if the plant species are the same of if the transgenes are alike, as variants of Bt-toxins which all may have an effect on herbivorous insects.

During the conference, the possibilities of risk assessment were discussed in great detail. In this context, it was clearly distinguished between issues which need to be studied, "*need-to-knows*", and facultative-ones, "*nice-to-knows*", which have no direct connection with transgenic crops. There was also agreement that the transgenes provide ideal markers for the tracing of all kinds of processes in population genetics. But this should not be a reason to deviate from the risk assessment discussion.

Monitoring must be hazard-oriented and useful for regulators

There is a need to obtain more information on outcrossing rates and the probability of transgene escape for various species, and therefore, a regionalised biogeographical assay including determination of local and regional natural hybridisation rates is essential. Furthermore, modelling may be very helpful, especially in the planning phase of monitoring when the crucial scenarios and parameters have to be selected. Scenarios resulting from modelling should always be compared with local data obtained by monitoring (and *vice versa*), as this process helps to understand complex ecosystems and also helps to discipline the thinking of ecologists. It was pointed out that farmers' observation networks can be of help for an early warning system as was shown in examples from the United States. In carefully assessed cases, monitoring of the spread of transgenes will have to be supported by genetic markers. The decision whether markers will be applied or not will depend on the plant species, the transgenes and the region of the release. In the future, monitoring also needs to include detection of environmental effects, e.g. to vegetation or to non-target organisms. Again the *case-by-case* principle is evident.

It was emphasised that there are many aspects of plant ecology where we only have limited knowledge, and this may restrict the use of transgenic plants. On the other hand, it was pointed out that our knowledge on crop plant behaviour is profound compared with that on other plant species, especially the wild relatives of well known crops. Furthermore, the use of transgenic plants will be of great value for science, especially to enhance knowledge on the principle of gene expression regulation and even more about the genomics and the function of many of the newly discovered genes.

In a general conclusion, the participants agreed on the principle that all risk assessment activities should be based on scientific grounds and that results should be validated step-by-step before they are accepted.

Acknowledgements

We want to sincerely thank to the following institutions for financial support:

FOEFL (Federal Office of Environment, Forests and Landscape, Bern, Switzerland)
Monsanto (Life Science Company, St. Louis, USA)
FOA (Federal Office of Agriculture, Bern, Switzerland)
SNF (Swiss National Foundation, Bern, Switzerland)
Novartis (Life Science Company, Basel, Switzerland)
Nestlé (Biotechnology, Vevey, Switzerland)
With this financial support, it was possible to issue an international invitation in order to get some of the best keynote speakers within their field of specialities. A few days after the symposium, we also culled a press conference, and a week later we made an effort to communicate results of the symposium to the public.

List of participants

ADAMS John
Department of Geography, UCL 26 Bedford Way WC1H OAP London UK
Tel. +44 171 387 70 50 Fax. +44 171 380 75 65 jadams@geography.ucl.ac.uk

AKEROYD John
Plant Talk P.O.Box 500, Kingston upon Thames KT2 5XB Surrey UK
Tel. +44 1747 871 5071 Fax. +44 1747 871 5071

AKIRA Hasebe
National Institute of Agrobiological Resources Kannondai 2-1-2 305 TSUKUBA Ibaraki Japan
Tel. +81 298 38 7452 Fax. +81 298 38 74 08 hasebe@abr.affrc.go.jp

AMMANN Klaus
Botanical Garden, University of Bern Altenbergrain 21 3013 BERN Switzerland
Tel. +41 31 631 49 37 Fax. +41 31 631 49 93 kammann@sgi.unibe.ch

BARBER Simon
OECD, Direction de l'Environnement 2 rue André Pascal, Cédex 16 F-75 775 Paris France
Tel. +33 1 45 24 16 78 Fax. +33 1 45 24 16 75 simon.barber@oecd.org

BAUMANN Peter
University of Bern Gesellschaftsstr. 31 CH-3012 Bern Switzerland
Tel. +41 31 301 80 54

BERETTA Isabella
Bundesamt für Bildung und Wissenschaft Hallwylstrasse 4 3003 Bern Switzerland
Tel. +41 31 322 84 17 Fax. +41 31 322 78 54 isabella.beretta@bbw.admin.ch

BERGMANS Hans
Committee on Genetic Modification Koningin Julianaplein 30-4B NL-2595 Den Haag AA
The Netherlands
Tel. +31 70 3045 932 Fax. +31 70 3045 939 cogem@solair1.inter.nl.net

BIERI Françoise
BICS Clarastrasse 13 CH-4058 Basel Switzerland
Tel. +41 61 690 93 21 Fax. +41 61 690 93 15 bieri@ubaclu.unibas.ch

BIGLER Franz
Swiss Federal Research Station for Agroecology Reckenholzstr. 191 CH-8046 Zürich
Switzerland

Tel. +41 1 377 72 35 Fax. +41 1 377 72 01 franz.bigler@fal.admin.ch

BINDLER Gregor N.
Philip Morris Europe, Research and Development c/o Fabriques de Tabac réunies S.A. 2003 Neuchâtel Switzerland
Tel. +41 32 888 76 18 Fax. +41 32 888 77 13

BODENMULLER Kurt
Internutrition Mainanstrasse 30, Postfach 1265 CH-8034 Zürich Switzerland
Tel. +41 1 383 12 00 Fax. +41 1 383 82 27 kb@wf-sdes.ch

BRAUN Paul
University Liebig Universität, Biometry und populations genetik Ludwigstr. 27 D-35390 Giessen Deutschland
Tel. +49 641 99 37 554 Fax. +49 641 99 37 549 gh80@agrar.uni-giessen.de

BURG Kornel
Austrian Research Centers A-2444 Seibersdorf Austria
Tel. +43 225 4780 3524 Fax. +43 2254 780 3653 burg@fzmail.arcs.ac.at

CAMP Karl-Heinz
Delley Samen und Pflanzenbau AG 1567 Delley Switzerland
Tel. +41 (0)26 677 90 20 camp.dsp@bluewin.ch

CAMPAGNA Norbert
20 rue de la Minière L-4839 Luxembourg, Luxembourg

COOPER J.I.
NERC IVEM Mansfield Road OX13SR Oxford UK
Tel. +44 01865 281 638 Fax. +44 01865 281 696 jic@mail.nerc-oxford.ac.uk

DALE Phil
John Innes Centre, Norwich Research Park Colney Lane NR4 7UH Norwich UK
Tel. +44 1603 452 571 Fax. + 44 1603 456 844

DAMGAARD Christian
National Environmental Research Institute, Dept. Terrestrial Ecology Vejlsovej 25 DK-8600 Silkeborg Denmark
Fax. +45 89 20 14 14 cfd@dmu.dk

DARMENCY Henri
Laboratoire de Malherbologie, INRA BV 1540 F-21034 Dijon Cédex France
Tel. +33 380 63 31 86 Fax. +33 380 63 32 62 henri.darmency@epoisses.inra.fr

DEN NIJS Hans
Institute for systematics & population Biology, University of Amsterdam Kruislaan 318 NL-

1098 Amsterdam SM The Netherlands
Tel. +31 20 525 7660 Fax. +31 20 525 7662 nijs@bio.uva.nl

DENYS Pascale
SES Europe Industriepark, Soldatenplein Z2, N° 15 3300 TIENEN BELGIUM
Tel. +32 16 80 82 11 Fax. +32 16 80 82 63

DORSCH-HASLER Karoline
Swiss Expert Commission for Biosafety c/o BUWAL CH-3003 Bern Switzerland
Tel. +41 31 323 03 55 Fax. +41 31 324 79 78 karoline.dorsch@buwal.admin.ch

ECOFFEY Danielle
Centre for Applied Studies in Int'l Negociations 11a Av. de la Paix CH-1202 Geneva Switzerland
Tel. +41 22 734 89 50 Fax. +41 22 73 64 44 ngocasin@iprolink.ch

ELVERS Heike
Institut für Zehrerforbildung Felix-Dahn-Str. 3 D-20 357 Hamburg Germany
Tel. +49 40 4212 722 Fax. +49 40 4212 2799

FELBER François
Institut de Botanique, Lab. de Phanérogamie, Université de Neuchâtel Chantemerle 18, C:P. 2 2007 Neuchâtel Switzerland
Tel. +41 32 718 23 39 Fax. +41 32 718 23 50 François.Felber@bota.uni.ch

FELDMANN Renata
Institute for Molecular Genetics, Biosafety research and consulting Johannes Gutenberg University D-55099 Mainz Germany
Tel. +49 61 31 39 52 24 Fax. +49 61 31 39 53 46 eschmidt@mzdmza.zdv.uni-mainz.de

FUCHS Marc
Cornell University, Dept. Plant Pathology New York State Agricultural Experiment Station 14456 GENEVA NY USA
Tel. +1 315 787 2351 Fax. +1 315 787 2389 mf13@cornell.edu

GADANI Ferruccio
Philip Morris Europe, Research and Development c/o Fabriques de Tabac réunies S.A. 2003 Neuchâtel Switzerland
Tel. +41 32 888 76 19 Fax. +41 32 888 77 13

GALLAND Nicole
Institut de Botanique Bâtiment de Biologie CH-1005 Lausanne Switzerland
Tel. +41 21 692 42 63 Fax. +41 21 692 41 05 nicole.galland@ibrg.inil.ch

GAY Philippe
Novartis Seeds AG R-1008.8.10 Ch-4002 Basel Switzerland

Tel. +41 61 697 44 57 Fax. +41 61 697 52 34 philippe.gay@seeds.novartis.com

GIDDINGS Glynies
Institute of Biological Sciences University of Wales Aberystwyth, Penglais Campus, Aberystwyth SY23 3EB Ceridigion UK
Tel. +44 1970 621 764 Fax. +44 1970 622 307

GORDON Allison
U.C.W. Aberystwyth Aberystwyth Ceredigion, Wales UK
Tel. 019 70 623 111 ext. Fax. 1784 01970 622 307 goa@aber.ac.uk

GUADAGNUOLO Roberto
Institut de Botanique, Lab. de Phanérogamie, Université de Neuchâtel Chantemerle 18, C.P.2 2007 Neuchâtel Switzerland
Tel. +41 32 718 23 70

HEMMER Wolfram
Bundesamt für Gesundheit Abteilung Lebensmittelwissenschaft, Sektion Mikrobiologieund Hygiene CH-3003 Bern Switzerland
Tel. +41 31 322 69 89 Fax. +41 31 322 95 74 wolfram.hemmer@bag.admin.ch

HESS Annatina
ETHZ, Institut für Terrestrische ökologie, Bodenbiologie Grabenstr. 3 8952 Schlieren Switzerland
Tel. +41 1 633 61 24 Fax. +41 1 633 11 22 hess@ito.umnw.ethz.ch

HILBECK Angelika
Swiss Federal Research Station for Agroecology and Agriculture Reckenholzstr. 191, Postfach CH-8046 Zürich Switzerland
Tel. +41 1 3777 410 Fax. +41 1 3777 201 angelika.hilbeck@fal.admin.ch

HIRSCHI Marlis
Gessellschaftstrasse 78 CH-3012 Bern Switzerland
Tel. +41 31 302 18 20

HOJLAND GRUNDTVIG Jan
National Forest and Nature Agency, Dept. Of Ecology Haraldsgade 53 DK-2100 Koebenhavn Denmark
Tel. +45 39 47 20 00 Fax. +45 39 27 98 99 jgh@sns.dk

HOUSEN Marnix
European Commission, JRC Ispra, Biotechnology & Environment TP 361 I-21020 Ispra VA Italy, Tel. +39 332 789 308 Fax. +39 332 785 483 marnix.housen@jrc.it

HUSBY Jan
Directorate for Nature Management Tungasletta 5 N-7005 Trondheim Norway

Fax. +47 739 154 33 Jan.Husby@)dnpost.md.dep.telemax.no

JACOT Yolande
Université de Neuchâtel, Laboratoire de Phanérogamie Chantemerle 22 CH -2007 Neuchâtel Switzerland
Tel. +41 32 931 90 08 Fax. +41 32 931 90 08 jacot@sgi.unibe.ch

JANS Beat
PRONATURA Postfach CH-4020 Basel Switzerland

JENAL-WANNER Ursula
Federal Office of Forests, Environment and Landscapes Hallwylstrasse 4 3003 Bern Switzerland
Tel. +41 31 323 03 50 ursula.jenal@buwal.admin.ch

KAREIVA Peter
University of Washington, Department of Zoology, Seattle, WA 98195 USA
Tel. +1 206 543 70 95 kareivap@zoology.washington.edu

KARLAGANIS Georg
Federal Office of Environment, Forests and Landscapes CH-3003 Bern Switzerland
Tel. +41 31 322 69 55 Fax. +41 31 324 79 78 georg.karlaganis@buwal.admin.ch

KELLER Martin
Eidg. Forschunganstalt für Agrarökologie und Landbau (FAL) Zürich Reckenholstr. 191 CH-8046 Zürich Switzerland
Tel. +41 1 377 74 46 Fax. +41 1 377 72 01 martin.keller@fal.admin.ch

KELLER-SENFTEN Julia
University of Bern, Botanical Institut Altenbergrain 21 CH-3013 Bern Switzerland
Tel. +41 31 631 49 27 Fax. +41 31 332 20 59

KJELLSSON Gösta
National Environmental Research Institute Veljsoweg 25, P.O. Box 314 8600 Silkeborg Denmark
Tel. +45 89 20 14 00 Fax. +45 89 20 14 14 gk@dmu.dk

KLINGER Terrie
Friday Harbor Labs, University of Washington 620 University Road, Friday Harbor, WA 98250, USA
Tel. +1 206 543 1484 Fax. +1 206 543 1273 tklinger@u.washington.edu

KOHLI Erich
BUWAL Abteilung Naturschutz, CH-3003 Bern Switzerland
Tel. +41 31 322 68 66 Fax. +41 31 323 03 47 erich.kohli@buwal.admin.ch

LACROIX Gill
Friends of the Earth Europe, Biotechnology programme 29, rue blanche B-1060 Brussels Belgium
Tel. +32 2 542 0182 Fax. +32 2 537 5596 100717.1155@compuserve.com

LANDSMANN Jörg
Federal Biological Research Centre for Agriculture and Forestry Institut for Biochemistry and Plant Virology, Messeweg 11-12 D-38104 Braunschweig Germany
Tel. +49 531 29 95 Fax. +49 531 299 30 06

LAUBER-FURST Susanne
FIAL Elfenstrasse 19 CH-3000 Bern 16 Switzerland
Tel. +41 31 352 11 88 Fax. +41 31 352 11 85

LEVIDOW Les
Centre for Technology Strategy Open University MK7 6AA Milton Keynes UK
Tel. +44 1908 653 672 Fax. +44 1908 652 175 l.levidow@open.ac.uk

MADSLUND Maya-Maria
Roskilde University, Department of Environment, Technology and Social Studies P.O. Box 260 DK-4000 Roskilde Denmark
Tel. +45 46 74 20 00 Fax. +45 46 74 30 41 mmm@teksam.ruc.dk

MAESCHLI Corinne
Fachstelle BATS Clarastrasse 13 CH-4058 Basel Switzerland
Tel. +41 61 690 93 14 Fax. +41 61 690 93 15

MALNOË Pia
Station Fédérale de Production Végétale 1260 Nyon Switzerland
Tel. +41 22 363 44 15 Fax. +41 22 362 13 25 pia.malnoe@rac.admin.ch

MESSMER Monika
Eidg. Forschunganstalt für Agrarökologie und Landbau (FAL) Zürich Reckenholstr. 191 CH-8046 Zürich Switzerland
Tel. +41 1 377 74 45 Fax. +41 1 377 72 01 monika.messmer@fal.admin.ch

MEYER Evelyn
Nestlé Biotechnology Coordination CH-1800 Vevey Switzerland
Tel. +41 21 924 22 52 Fax. +41 21 924 28 51 evelyn.meyer@nestle.com

MEYER Rolf
Büro für Technikfolgen-Abschätzung beim Deutschen Bundestag Rheinweg 121 D-53129 Bonn Germany
Tel. +49 228 233 583 Fax. +49 228 233 755 buero@tab.fzk.de

MICHNA Kristin
Natur Media GmbH Belfortstrasse 8 D-81667 München Germany
Tel. +49 89 458 727 29 Fax. +49 89 458 727 24 kristin@natur.de

MINOL Klaus
Institut for Biochemie, TU Darmstadt Petersenstr. 22 D-64287 Darmstadt Germany
Tel. +49 6151 162 175 Fax. +49 6151 165 175

MUELLER Werner
Institute of Organic Farming, University of Agricultural Sciences (BOKU) Gregor-Mendelstr. 33, A-1190 Vienna, AUSTRIA
Tel. +43 1 47 654 37 58 Fax. +43 1 47 654 37 92 wmueller@edv1.boku.ac.at

MUSTER Matthias
Institute of Plant Physiology, Department of Molecular Biology Altenbergrain 21 CH-3013 Berne Switzerland
Tel. +41 31 631 37 77 Fax. +41 31 332 20 59 matthias.muster@pfp.unibe.ch

NAYBERG Helle
Danish Environmental Protection Agency Strandgade 29 DK-1401 Copenhagen K Denmark
Tel. +45 32 66 01 00/362 Fax. +45 32 66 04 79 hna@mst.dk

NEEMANN Gerd
Büro für Landschaftsökologie und Umweltstudien (BLaU) Göttingen Wiesenstrasse 8 D-37073 Göttingen Germany
Tel. +49 551 703 435 Fax. +49 551 703 536 BLaU.@t-online.de

NICKSON Thomas E.
Monsanto, Life Sciences 700 Chesterfield Parkway North, St. Louis, MO 63198, USA
Tel. +1 314 737 6688 Fax. +1 314 737 6189 Thomas.Nickson@monsanto.com

NOEH Ingrid
Umweltbundesamt Berlin, Fachgebietslehrerin Vollzug GenTG Seeckstr. 6-10, Postfach 33 00 22 D-14191 Berlin Germany
Tel. +49 3089 0332 50 Fax. +49 3089 0332 52 ingrid.noeh@uba.de

NORRIS Carol
National Institute of Agricultural Botany Huntingdon Road CB3 OLE Cambridge UK
Tel. +44 12 23 233 258

OOSTERMEIJER J. Gerard B.
Institute for Systematics & Population Biology, Hugo de Vries Laboratory, University of Amsterdam Kruislaan 318 NL-1098 Amsterdam SM The Netherlands
Tel. +31 20 525 76 60 Fax. +31 20 525 76 62

POHL Matthias
RWTH Worringerweg 1 D-52056 Aachen Germany
Tel. +49 241 80 66 76 Fax. +49 241 88 88 182 pohl-orf@rwth-aachen.de

POIRIER Yves
Université de Lausanne, Institut de Biologie et Physiologie Végétale Bâtiment de Biologie CH-1015 Lausanne Switzerland
Tel. +41 21 692 41 90 Fax. +41 21 692 41 95 Yves.Poirier@ibpv.unil.ch

POULSEN Gitte Silberg
Danish Environmental Protection Agency Strandgade 29 DK-1401 Copenhagen, Denmark
Tel. +45 32 60 03 71 Fax. +45 32 60 04 62 gsp@mst.dk

PYTHOUD François
BUWAL Hallwylstrasse 4 3003 BERN Switzerland
Tel. +41 31 322 93 95 Fax. +41 31 324 79 78

RAPS Andrea
FAL Reckenholz Reckenholzstr. 191 CH-8046 Zürich Switzerland
Tel. +41 1 377 74 30 Fax. +41 1 377 72 01

RAYBOULD Alan
Centre for Ecology and Hydrology Furzebrook Research Station, Furzebrook Road BH20 5AS Wareham Dorset UK
Tel. +44 1929 551 518 Fax. +44 1929 551 087 afr@wpo.nerc.ac.uk

REBOUD Xavier
Laboratoire de Malherbologie, INRA BV 1540 F-21034 Dijon Cédex France
Tel. +33 380 63 31 86 Fax. +33 380 63 32 62

RENCKENS Suzy
Institute of Public Health, Biosafety and Biotechnology J. Wytsmanstraat 14 B-1040 Brussels Belgium
Tel. +32 642 53 37 Fax. +32 2 642 52 92 srencken@sbb.ihe.be

RICHTER Hannes
University of Lausanne IBPV Bâtiment de Biologie CH-1015 Lausanne Switzerland
Tel. +41 21 692 42 38 Fax. +41 21 692 41 95 hrichter@ibpv.unil.ch

ROSSI Luca
Philip Morris Europe, Research and Development c/o Fabriques de Tabac réunies S.A. 2003 Neuchâtel Switzerland
Tel. +41 32 888 76 50 Fax. +41 32 888 77 13

RUDENKO Larisa
ENVIRON International Corp. 4350 Fairfax Drive, Arlington, VA 22203, USA

Tel. +1 703 516 23 02 Fax. +1 703 516 23 93 lrudenko@environcorp.com

RUFENER AL MAZYAD Pia
Botanical Garden, University of Bern Altenbergrain 21 CH-3013 Bern Switzerland
Tel. +41 31 631 37 67 Fax. +41 31 631 49 93

RUOHONEN-LEHTO Marja
Finnish Environment Institute P.O. Box 140 Fin-00251 Helsinki Finnland
Fax. +358 9 4030 0591 marja.ruohonen-lehto@vyh.fi

SACHSE Gabriele
Bioalliance GmbH Steinle strasse 6 D-60596 Frankfurt Germany
Tel. +49 69 631 464 0 Fax. +49 69 631 464 70

SAVOVA-BIANCHI Dessislava
Institut de Botanique, Lab. de Phanérogamie, Université de Neuchâtel Chantemerle 18, C.P.2
2007 Neuchâtel Switzerland
Tel. +41 32 718 23 67 Fax. +41 32 718 30 01 Dessislava. Savova@bota.unine.ch

SCHENKELAARS Piet
Schenkelaars Biotechnology Consultancy, Bio Sciencepark, Niels Bohrweg 11–13, NL-2333
CA Leiden, The Netherlands
Tel: +31 71 523 50 89 Fax: +31 71 523 50 90 pschenkelaars.sbcbiotech@wxs.nl

SCHUETZ Otto
öVAF, Austrian Association for Agricultural Research Kleine Sperlgasse 1-37 A-1020 Wien,
Austria
Tel. +43 1 214 59 03 Fax. +43 1 214 59 03/9 oevaf-wien@oevaf.at

SCHULTE Elisabeth
BATS Clarastrasse 13 4058 Basel Switzerland
Tel. +41 61 690 93 12 Fax. +41 61 690 93 15

SCHWEIZER Gaby
FACTS, Das Schweizer Nachrichtenmagazin Postfach, CH-8034 Zürich, Switzerland
Tel. +41 1 389 32 88 Fax. +41 38932 70 gschweizer@facts.ch

SEITER Andreas
Novartis Deutschland GmbH, Öflinger Strasse 44 D-79664 Wehr/Baden Germany
Tel. +49 77 62 82 27 03 Fax. +49 77 62 35 95 andreas.seiter@group.novartis.com

SHI-RONG Jia
Biotechnology Research Center, Chinese Academy of Agricultural Sciences 30 Baishigiao
Road 100081 Beijing China
Tel. +86 10 621 72 803 Fax. +86 10 621 72 803 jiasr@public3.bta.net.cn

SIMONSEN Vibeke
National Environmental Research Institute Veljsoweg 25, P.O. Box 314 8600 Silkeborg Denmark
Tel. +45 89 20 14 00 Fax. +45 89 20 14 14 vs@dmu.dk

SKORUPINSKI Barbara
Institut für Sozialethik, Universität Zürich Zollikerstrasse 117 CH Zürich Switzerland
Tel. +41 1 385 45 17 Fax. +41 1 385 45 08 bask@sozethik.unizh.ch

SOLBE John
Unilever Research Quarry Road East, Bebington LB3 3JW Wirrat UK
Tel. +44 151 471 32 05 Fax. +44 151 471 18 47 john.f.solbe@unilever.com

SOTH Jens
Environmental Protection Encouragement Agency Feldstrasse 36 D-20357 Hamburg Germany
Tel. +49 40 439 20 81 Fax. +49 40 439 20 85

STOLL Peter
University of Bern, Geobotanisches Institut Altenbergrain 21 CH-3013 Bern Switzerland
Tel. +41 31 631 49 49 Fax. +41 31 332 20 59 stoll@sgi.unibe.ch

STORENG Anne Britt
Directory for Nature Management Tungasletta 2 N-7005 Trondheim Norway
Tel. +47 735 798 00 Fax. +47 739 154 33 Anne-Britt.storeng@dnpost.md.dep.teiemax.no

SWEET Jeremy
National Institute of Agricultural Botany Huntingam Road CB 3 OLE Cambridge UK
Tel. +44 0 1223 276 381 Fax. +44 1 12 23 324 394 J.B.Sweet@pvs.maff.gov.uk

THYSSEN Marco
Monsanto 270-272 Av. de Tervuren B-1150 Bruxelles Belgium
Tel. +32 2 776 47 25 Fax. +32 2 776 44 06 marco.thyssen@monsanto.com

TOET Dirk
Nestec Ltd, Biotechnology Co-ordination Manager Av. Nestlé 55 CH-1800 Vevey Switzerland
Tel. +41 21 924 31 49 Fax. +41 21 924 28 51

TOMMERAS Bjorn Age
Norwegian Institute for Nature Research Tunglasletta 2 N-7005 Trondheim Norway
Tel. +47 73 58 06 21 Fax. +47 73 91 54 33 bjorn.age.tommeras@nina.nina.no

TUFTO Jarle
ICAPB, University of Edinburgh, King's Buildings, West Mains Road, Edinburgh EH9 3JT, UK

Tel: +44 131 650 5508 jarle.tufto@ed.ac.uk

VAN DE GRAAF Ronald
RIVM P.O. Box 1 3720 Bilthoven BA The Netherlands
Tel. +31 30 274 41 76 Fax. +31 30 274 44 61 CSR.990@RIVM.nl

VISHNEVSKA Ludmila
Scientific secretary and data base manager of National Botanic Gardens LV-2169 Salaspils-1 Latvia
Tel. +371 2 944 974 Fax. +371 790 12 50 nbd@com.latnet.lv

WHITE James
US Department of Agriculture Unit 147, 4700 River Road Riverdale MD 20737 USA
Tel. +1 301 734 59 40 Fax. +1 301 734 86 69 jwhite@aphis.usda.gov

WICKE Gisela
Koordinierungsstelle Volksbegehren und Landesverband Naturschutzbund Deutschland Erichshof D-30989 Gehrden Germany
Tel. +49 511 394 34 44 Fax. +49 511 394 34 45

WURMBACH Peter
Robert Koch-Institut Bundesinstitut für Infektionskrankheiten und nicht übertragbare Krankheiten, Fachbereich Genetik, Gentechnik, Postfach 34 D-13161 Berlin Germany
Tel. +49 30 454 730 18 Fax. +49 30 454 730 60

XIONG BAO Li
China National Center for Biotechnology Development B7# Zao Jun Miao 100081 Beijing China
Tel. +86 10 622 51 984 Fax. +86 10 622 54 106

ZADOKS Jan Carel
Committee on Genetic Modification Hendrikweg 31 NL-6703 Wageningen AV The Netherlands
Tel. +31 317 414 478 Fax. +31 317 423 614 jczadoks@user.diva.nl

ZESIGER Matthias
University of Bern Rainstrasse 29 CH-2503 Biel Switzerland
Tel. +41 32 365 75 89

ZIEGLER Ruediger
Union Deutsche Lebensmittelwerke GmbH Dammturwall 15 D-20355 Hamburg Germany
Tel. +49 40 349 311 64 Fax. +49 40 349 314 53

Subject index

The page numbers refer to the first page of each chapter in which a term is discussed.

MIX
Papier aus verantwortungsvollen Quellen
Paper from responsible sources
FSC® C105338

If you have any concerns about our products,
you can contact us on
ProductSafety@springernature.com

In case Publisher is established outside the EU,
the EU authorized representative is:
Springer Nature Customer Service Center GmbH
Europaplatz 3, 69115 Heidelberg, Germany

Printed by Libri Plureos GmbH
in Hamburg, Germany